Triumph Trident
BSA Rocket 3
Owners
Workshop
Manual

By Frank Meek

Models covered
741 cc Triumph T150 Trident. 1969 to 1972
741 cc Triumph T150V Trident. 1972 to 1975
741 cc BSA A75 Rocket 3. 1969 to 1972

ISBN 978 0 85696 136 6

© Haynes Publishing 2002

All rights reserved. No part of this book may be reproduced or transmitted in
any form or by any means, electronic or mechanical, including photocopying,
recording or by any information storage or retrieval system, without permission
in writing from the copyright holder.

ABCDE
FGHIJ
K

2

Printed in the UK   *(136 – 5AB9)*

Haynes Publishing Group
Sparkford Nr Yeovil
Somerset BA22 7JJ England

Haynes Publications, Inc
859 Lawrence Drive
Newbury Park
California 91320 USA

Acknowledgements

Our thanks are due to Norton Triumph International Limited for their assistance. Brian Horsfall and Martin Penny gave the necessary assistance with the overhaul and devised ingenious methods for overcoming the lack of service tools. Les Brazier took the photographs that accompany the text. Jeff Clew edited the text.

We are especially grateful to the Triumph Owners M.C.C. and in particular to Clifford Garside, who loaned us the 1972 five-speed "Trident" on which most of the workshop sequences are based. We would also like to acknowledge the help of the Avon Rubber Company, who kindly supplied illustrations and advice about tyre fitting, Amal Limited for the use of their carburettor illustrations, Boyer Racing for advice about the use of their transistorised ignition system.

The cover photograph was arranged through the courtesy of Vincent and Jerrom Ltd of Taunton.

About this manual

The author of this manual has the conviction that the only way in which a meaningful and easy-to-follow text can be written is to participate in the work himself.

The demonstration machine is not new and a machine which had travelled a good mileage and was in need of repair was chosen. This was so that the conditions encountered would be typical of those encountered by the average rider/owner.

Unless specially mentioned Triumph or B.S.A. tools have not been used: There are invariably alternative means of slackening or removing some vital component when service tools are not available. Risk of damage is to be avoided at all cost.

Each of the eight Chapters is divided into numbered Sections. Within the sections are numbered paragraphs. Cross-reference throughout the manual is quite straightforward and logical. For example, when reference is made 'See Section 6.2' it means section 6, paragraph 2 in the same chapter. If another chapter were meant, the reference would read 'See Chapter 2, section 6.2'. All photographs are captioned with a section/paragraph number to which they refer, and are always relevant to the chapter text adjacent.

Figure numbers (usually line illustrations) appear in numerical order, within a given chapter. Fig. 1.1. therefore refers to the first figure in Chapter 1. Left hand and right hand descriptions of the machines and their component parts refer to the left and right when the rider is seated, facing forward.

Motorcycle manufacturers continually make changes to specifications and recommendations, and these, when notified, are incorporated into our manuals at the earliest opportunity.

Whilst every care is taken to ensure that the information in this manual is correct no liability can be accepted by the authors or publishers for loss, damage or injury, caused by any errors in or omissions from the information given.

Introduction
to the Triumph 750cc Trident and BSA Rocket 3

The Triumph "Trident" and the BSA "Rocket 3" were first introduced to the UK market during April 1969. It had been known for some considerable while that both Triumph and BSA were experimenting with a three cylinder model since test machines, devoid of all identification, had been seen on the roads. There was certainly a demand for both models, the choice depending on the purchasers' allegiance to the particular name. It was not long before the three cylinder models appeared in racing events, especially the works prepared models ridden by Percy Tait, John Cooper and Tony Jefferies, to name but three of the works sponsored riders. Even the great Mike Hailwood rode a BSA three as a member of the British Team in one of the Transatlantic Match Races.

Although there is a pronounced basic similarity between the Triumph and BSA threes, it is the Trident that somehow achieved the greatest racing successes. A Trident has won the 750 cc Production TT for five consecutive years, a quite remarkable achievement at the very time when machines of Japanese origin have almost completely monopolised racing.

Today, the production of Tridents has been transferred from Meriden to the old BSA factory at Armoury Road, Birmingham. The Rocket 3 became obsolete during 1972 when the production of BSA motor cycles ceased completely. The latest models boast a five-speed gearbox, disc brakes, electric starter and a three into one exhaust system, in a successful bid to keep pace with current design trends.

They have also been extensively re-styled, to present a more aggressive appearance.

Contents

Triumph Trident T150 1970 (ex-police model)

1972 BSA Rocker 3

1974 Triumph 750 cc 'Trident'

1972 T150 Trident (US export)

Ordering spare parts

When ordering spare parts for any of the Triumph or BSA three cylinder models, it is advisable to deal direct with an official Triumph agent who will be able to supply many of the items ex-stock. Parts cannot be obtained direct from Norton-Triumph International Limited; all orders must be routed through an approved agent, even if the parts required are not held in stock.

Always quote the engine and frame numbers in full. Include any letters before or after the number itself. The frame number will be found stamped on the left hand front down tube, adjacent to the steering head. The engine number is stamped on the left hand crankcase, immediately below the base of the cylinder barrel.

Use only parts of genuine Triumph manufacture. Pattern parts are available but in many instances they will have an adverse effect on performance and/or reliability. Some complete units are available on a 'service exchange' basis, affording an economic method of repair without having to wait for parts to be reconditioned. Details of the parts available, which include petrol tanks, front forks, front and rear frames, clutch plates, brake shoes etc can be obtained from any Triumph agent. It follows that the parts to be exchanged must be acceptable before factory reconditioned replacements can be supplied.

Location of engine number

Location of frame number

Routine maintenance and capacities data

ENGINE (Fuel)	15.9 litres (3.5/4.2 Imp/US gallons)
OIL TANK	3.4 litres (6.0/7.2 Imp/US pints) SAE 20W/50 engine oil
GEARBOX	710 cc (1.25/1.50 Imp/US pints) EP90 hypoid gear oil
FRONT FORKS (Each leg)	190 cc (6.6/6.4 Imp/US fl oz) SAE 20 fork oil
PRIMARY CHAINCASE (Initial fill)	355 cc (12.4/12.0 Imp/US fl oz) SAE 20W/50 engine oil
CONTACT BREAKER GAP	0.014 - 0.016 inches 0.35 - 0.40 mm
SPARK PLUG GAP	0.025 inches 0.625 mm

TYRE PRESSURES*	Front 26 lbs/sq inch
	            1.828 kg/sq cm
	Rear 28 lbs/sq inch
	            1.97 kg/sq cm

*Note 1   Suitable for a 12 stone (76 kg) rider. For pillion passenger increase by 6 lb/sq inch
*Note 2   Speeds over 100 mph use 32 lb/sq inch (2.25 kg/sq cm) front
                33 lb/sq inch (2.32 kg/sq cm) rear

Routine maintenance

The need for weekly routine maintenance is something which cannot be over-emphasised. As soon as the machine is ridden every part is under some form of stress and it must be treated almost as though it were a living thing. If the machine is not used regularly, routine maintenance must still be carried out to prevent any components falling into decay.

Whilst keeping the machine in good condition, routine maintenance can prove invaluable as a kind of early warning system against failures of all kinds, mechanical and electrical. Charts and diagrams are shown to help and guide you, with dates and mileages. It should be remembered that the intervals between various maintenance tasks serve only as a guide. As the machine gets older, or is perhaps subjected to harsh conditions, it is advisable to reduce the period between each check.

Some tasks are described in detail, but where they are not mentioned fully as routine maintenance items, they will be found elsewhere in the text under their appropriate chapters. No special tools are necessary for normal maintenance jobs, but those in the machine's tool kit coupled perhaps with those in the average garage at home should prove quite adequate for the task.

As well as keeping the bike in good condition mechanically, routine maintenance will pay dividends when you sell it. The Service Charts give no mention of brakes, which must be religiously checked and adjusted at all times. You are inclined to remember your brakes only when you need them, and if they have been neglected this may prove too late.

Consult the Owners Handbook on the routine maintenance. Later models may have modifications that require changes to the maintenance schedule.

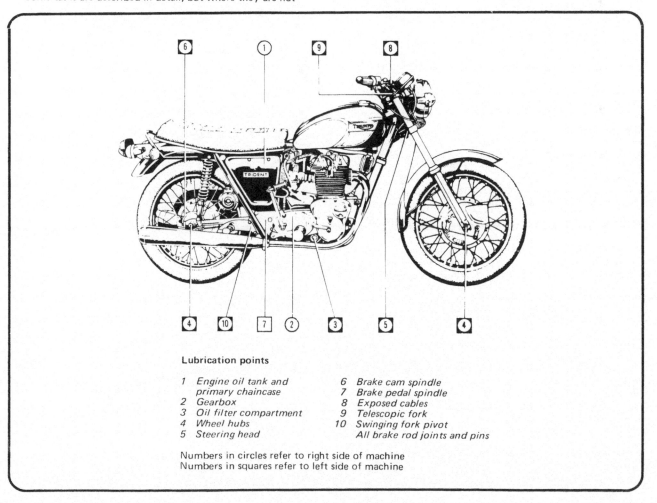

Lubrication points

1	Engine oil tank and primary chaincase	6	Brake cam spindle
2	Gearbox	7	Brake pedal spindle
3	Oil filter compartment	8	Exposed cables
4	Wheel hubs	9	Telescopic fork
5	Steering head	10	Swinging fork pivot
			All brake rod joints and pins

Numbers in circles refer to right side of machine
Numbers in squares refer to left side of machine

Weekly or every 250 miles (400 km)

Check level in oil tank and top up if necessary
Check level in primary chaincase
Lubricate rear chain
Check battery acid level and top up with distilled water if necessary
Check tyre pressures

Monthly or every 1000 miles (1600 km)

Change oil in primary chaincase
Lubricate all cables including brakes*
Grease swinging arm fork pivot
Remove, clean and lubricate final drive chain and re-adjust when refitted
Check and tighten nuts and bolts if necessary
Adjust tension of primary chain

Six weekly or every 1500 miles (2400 km)

Change oil in gearbox, also in front forks
Apply a light smear of grease to contact breaker cam
Check adjustment of steering head bearings

Three monthly or every 3000 miles (4800 km)

Change the engine oil and clean all filters in the lubrication system
Adjust points and clean and lubricate contact breaker heel and check ignition timing
Grease brake pedal spindle
Clean plugs and regap
Clean carburettors
Check adjustment of valve operating mechanism

Six monthly or every 6000 miles (9600 km)

Check oil level in gearbox
Examine front forks for oil leakage
Check valve clearances
Clean air filters and clean and adjust carburettor(s)

Yearly or every 12,000 miles (19,200 km)

Grease wheel bearings and steering head bearings

The 12,000 mile (19,200 km) service will mean a reasonable amount of dismantling (all details given in appropriate chapters). It should be noted that even when six-monthly and yearly maintenance has to be undertaken, the weekly, monthly, six-weekly and three-monthly services must be carried out in between. There is no part of a motor cycle's life when any routine maintenance tasks can be ignored or left out.

A few other items to be checked regularly:

The electrical system must be in good working order at all times.
Tyres must be maintained at the correct pressure and adequately treaded for the job (ie solo or sidecar). They can be cut in a second so it is advisable to check regularly for signs of cracks or splits, bulging or uneven wear

All these things make common sense and are life-savers — someone else's and yours.

Adjusting the front brake (drum brake models only)

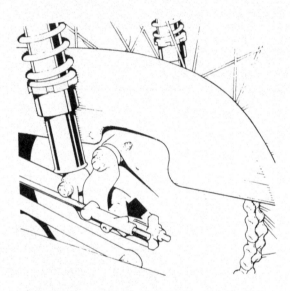

Adjusting the rear brake (drum brake models only)

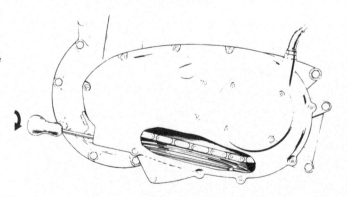

Adjusting the primary chain tension

Chapter 1 Engine

Contents

Specifications

Triumph Trident T150 and BSA Rocket 3
Basic Details

Bore and stroke mm...	...	...	...	...	...	...	67 x 70
Bore and stroke ins ...	...	...	...	...	...	...	2.67 x 2.751
Cubic capacity cc ...	...	...	...	...	...	...	741
Cubic capacity cub ins	...	...	...	...	...	...	45
Compression ratio ...	...	...	...	...	...	...	9.5 : 1
							(9 : 1 earlier models)

Crankshaft

Type ...	...	...	...	...	...	...	...	En 16 B Hardened and tempered stamping - one piece
Main bearing (Drive side) type	...	...	...	...	...	Hoffman MS11
Size ins ...	...	...	...	...	...	...	1 1/8 x 2 13/16 x 13/16 caged ball
Size mm ...	...	...	...	...	...	...	28.58 x 71.43 x 20.63
Main bearing (timing side) type	...	...	...	...	...	Hoffman R125
Size ins ...	...	...	...	...	...	...	1 1/8 x 2 13/16 x 13/16 roller
Size mm ...	...	...	...	...	...	...	28.58 x 71.43 x 20.62

Main bearing (centre)
   Running clearance

ins	.0005 - .0022
mm	.0127 - .05588

Right hand main bearing housing

Diameter ins	2.8110 - 2.8095
Diameter mm	71.3994 - 71 - 3613

Right hand main bearing journal

Diameter ins	1.1248 - 1.1245
Diameter mm	28.5699 - 28.5623

Centre main bearing housing

Diameter ins	2.0630 - 2.0625
Diameter mm	52.4002 - 52.3875

Centre main bearing journal

Diameter ins	1.9170 - 1.9175
Diameter mm	48.6918 - 48.7045
Left main bearing housing diameter ins	2.0447 - 2.0457
Left main bearing housing diameter mm	51.9344 - 51.9608
Left main bearing journal diameter ins	0.9843 - 0.9840
Left main bearing journal diameter mm	25.0012 - 24.9936
Big end journal diameter ins	1.6240 - 1.6235
Big end journal diameter mm	41.2496 - 41.2369
Minimum regrind diameter ins	1.6200 - 1.6185
Minimum regrind diameter mm	41.148 - 41.1099
Crankshaft end float ins	.0015 - .0145
Crankshaft end float mm	.038 - .368

Connecting rods

Material	Alloy "H" section RR56
Length (centres) ins	5.751 - 5.749
Length (centres) mm	146.075 - 146.024
Big end bearing type	White metal steel backed
Rod side clearance ins	0.013 - 0.019
Rod side clearance mm	0.3302 - 0.4826
Bearing diametrical clearance ins	0.0005 - 0.0020 min
Bearing diametrical clearance mm	0.0127 - 0.0508

Gudgeon pins

Material	High tensile steel
Fit in small end ins	0.0005 - 0.0011
Fit in small end mm	0.0127 - 0.0279
Diameter ins	0.6883 - 0.6885
Diameter mm	17.4828 - 17.4880
Length ins	2.250 - 2.235
Length mm	57.150 - 56.769

Cylinder block

Material	Aluminium alloy with Austenitic steel. liner
Bo re size (Standard) ins	2.6368 - 2.6363
Bore size (Standard) mm	66.9747 - 66.962

Rebore sizes

Piston size

(Standard) ins	2.6368 - 2.6363
(Standard) mm	66.975 - 66.962
+ 0.010 (0.254 mm) ins	2.6468 - 2.6463
+ 0.010 (0.254 mm) mm	67.229 - 67.215
+ 0.020 (0.508 mm) ins	2.6568 - 2.6563
+ 0.020 (0.508 mm) mm	67.483 - 67.470
+ 0.040 (1.016 mm) ins	2.6768 - 2.6763
+ 0.040 (1.016 mm) mm	67.990 - 67.980

Regrind sizes

Big end bearings

Shell bearing marking suitable crankshaft size

	Inches	mm
Standard	1.6235	41.237
	1.6240	41.250

Undersize 0.010	...	...	...	...	...	...	...	1.6135	40.985
								1.6140	40.996
Undersize 0.020	...	...	...	...	...	...	...	1.6035	40.729
								1.6040	40.742
Undersize 0.030	...	...	...	...	...	...	...	1.5935	40.475
								1.5940	40.488
Undersize 0.040	...	...	...	...	...	...	...	1.5835	40.221
								1.5840	40.234

Centre main bearings

Standard	...	...	...	...	...	...	...	1.9170	48.692
								1.9175	48.705
Undersize 0.010	...	...	...	...	...	...	...	1.9070	48.438
								1.9075	48.451
Undersize 0.020	...	...	...	...	...	...	...	1.8970	48.184
								1.8975	48.197
Undersize 0.030	...	...	...	...	...	...	...	1.8870	47.930
								1.8875	47.943
Undersize 0.040	...	...	...	...	...	...	...	1.8770	47.676
								1.8775	47.689

Cylinder head

Material	...	...	...	...	...	...	...	Aluminium alloy die casting
Inlet port size ins		...	...	...	...	...	...	1.0 diameter
Inlet port size mm		...	...	...	...	...	...	25.4
Exhaust port size ins		...	...	...	...	...	...	1¼
Exhaust port size mm		...	...	...	...	...	...	31.75
Valve seating material		...	...	...	...	...	...	Cast iron

Valves

Stem diameter inlet ins	...	...	...	...	...	...	0.3100 - 0.3095
Stem diameter inlet mm	...	...	...	...	...	...	7.8740 - 7.8613
Stem diameter exhaust ins	...	...	...	...	...	...	0.3095 - 0.3090
Stem diameter exhaust mm	...	...	...	...	...	...	7.8613 - 7.8495
Head diameter inlet ins	...	...	...	...	...	...	1.534 - 1.528
Head diameter inlet mm	...	...	...	...	...	...	38.9638 - 38.812
Head diameter exhaust ins	...	...	...	...	...	...	1.315 - 1.309
Head diameter exhaust mm	...	...	...	...	...	...	33.401 - 33.2486
Exhaust valve material	...	...	...	...	...	...	24/4 'N' Heat treated

Valve guides

Material	...	...	...	...	...	...	...	Hidural 5
Bore diameter (inlet and exhaust) ins	...	...	...	...	0.3115 - 0.3110
Bore diameter (inlet and exhaust) mm	...	...	...	...	7.9121 - 7.8994
Outside diameter (inlet and exhaust) ins	...	...	...	...	0.5005 - 0.5010
Outside diameter (inlet and exhaust) mm	...	...	...	...	12.7127 - 12.7254
Length : (Inlet and Exhaust) ins	...	...	...	...	...	1.875
Length : (Inlet and Exhaust) mm	...	...	...	...	...	47.625

Valve springs (Red and White)

Free length inner ins	...	...	...	...	...	...	1.468
Free length inner mm	...	...	...	...	...	...	37.2872
Free length outer ins	...	...	...	...	...	...	1.600
Free length outer mm	...	...	...	...	...	...	40.64
Number of coils inner	...	...	...	...	...	...	6
Number of coils outer	...	...	...	...	...	...	5½
Total fitted load							
Valve open: inner lbs	...	...	...	...	...	...	82
Valve open: inner kgm	...	...	...	...	...	...	37.228
Valve open: outer lbs	...	...	...	...	...	...	115
Valve open: outer kgm	...	...	...	...	...	...	51.31
Valve closed: inner lbs	...	...	...	...	...	...	37 - 40
Valve closed: inner kgm	...	...	...	...	...	...	16.798 - 18.144
Valve closed: outer lbs	...	...	...	...	...	...	45 - 53
Valve closed: outer kgm	...	...	...	...	...	...	21.792 - 24.062

Valve timing

(All clearances at nil for checking. Valve lift to be measured at
T.D.C. with cold engine)

Valve lift: inlet ins	...	...	...	...	...	...	0.152

Valve lift: inlet mm ...	3.86
Valve lift: exhaust ins	0.146
Valve lift: exhaust mm	3.71

Rockers

Material	EN 33 stamping NI. CH.
Bore diameter ins	0.5002 - 0.5012
Bore diameter mm	12.7051 - 12.7305
Rocker spindle diameter ins	0.4990 - 0.4995
Rocker spindle diameter mm	12.6746 - 12.6873
Tappet clearance (cold) inlet ins	.006
Tappet clearance (cold) inlet mm	0.1524
Tappet clearance (cold) exhaust ins	.008
Tappet clearance (cold) exhaust mm	0.2032

Camshafts

Journal diameters ins	1.0615 - 1.0605
Journal diameters mm	26.9621 - 26.9367
Diameteral clearances ins	.0005 - .0020
Diametral clearances mm	.0127 - .0508
End float ins	.007 - .014
End float mm	0.178 - 0.356
Cam lift: inlet and exhaust ins	0.3045
Cam lift: inlet and edhaust mm	7.7343
Base circle diameter ins	.812
Base circle diameter mm	20.6248

Tappets

Material	EN 32 B Stellite tip
Tip radius ins	1.125
Tip radius mm	28.575
Tappet diameter ins	0.3115 - 0.3110
Tappet diameter mm	7.9121 - 7.8994
Clearance in guide block ins	0.0005 - .0015
Clearance in guide block mm	0.0127 - .0381

Tappet guide block

Diameter of bores ins	0.3125 - 0.3120
Diameter of bores mm	7.9375 - 7.9248
Outside diameter ins	1.153 - 1.148
Outside diameter mm	29.2862 - 29.1592
Interference fit in cylinder block ins	0.0027 - 0.0082
Interference fit in cylinder block mm	0.06858 - 0.20

Rocker spindle bushes

Bush D/S Bore diameter ins	0.497 - 0.498
Bush D/S Bore diameter mm	12.624 - 12.649
Bush D/S Outside diameter ins	0.6260 - 0.6265
Bush D/S Outside diameter mm	15.9004 - 15.913
Bush T/S Bore diameter ins	0.375 - 0.374
Bush T/S Bore diameter mm	9.525 - 9.4996
Bush T/S Outside diameter ins	0.501 - 0.502
Bush T/S Outside diameter mm	12.725 - 12.751

Timing gears

Inlet and exhaust camshaft pinions	
Number of teeth	50
Interference fit on camshaft ins	0.000 - 0.001
Interference fit on camshaft mm	0.000 - 0.0254
Intermediate timing gear	
Number of teeth	42
Bore diameter ins	0.5618 - 0.5625
Bore diameter mm	14.2697 - 14.2875
Needle roller ins	11/16 x 7/8 x 5/8
Needle roller mm	17.46 x 22.225 x 15.87
Intermediate wheel spindle	
Diameter ins	0.6888 - 0.6885
Diameter mm	17.4955 - 17.4879
Crankcase pinion	
Number of teeth	25
Fit on crankcase + ins	0.0003
Fit on crankcase + mm	0.00762
Fit on crankcase − ins	0.0005
Fit on crankcase − mm	0.0127

Pistons

Material	"Lo-ex" Aluminium alloy die casting
Clearance top of skirt ins	0.0056 - 0.0035

Clearance top of skirt mm ...	...	...	...	...	...		0.42 - 0.089
Bottom of skirt ins ...	...	...	...	...	...		0.0033 - 0.0018
Bottom of skirt mm	...	...	...	...	...		0.84 - 0.0457

Piston rings

Material compression top and centre	...	...	...	...			Cast iron HG 10
Scraper "apex" ...	...	...	...	...	...		Chrome plated steel
Compression rings - tapered							
Width mm ...	...	...	...	...	...	...	2.729 - 2.577
Thickness ins ...	...	...	...	...	...	...	0.0625 - 0.0615
Thickness mm ...	...	...	...	...	...	...	1.5875 - 1.5621
Fitted gap ins ...	...	...	...	...	...	...	0.009 - 0.013
Fitted gap mm ...	...	...	...	...	...	...	0.2286 - 0.3302
Clearance in groove ins ...	...	...	...	...	...	...	0.0035 - 0.0015
Clearance in groove mm	...	...	...	...	...	...	0.89 - 0.038
Oil control ring							
Width mm ...	...	...	...	...	...	...	2.729 - 2.577
Thickness ins ...	...	...	...	...	...	...	0.125 - 0.124
Thickness mm ...	...	...	...	...	...	...	3.175 - 3.1496
Fitted gap ins ...	...	...	...	...	...	...	0.010 - 0.040
Fitted gap mm ...	...	...	...	...	...	...	0.254 - 1.016
Clearance in groove ins ...	...	...	...	...	...	...	0.0105 - 0.0065
Clearance in groove mm	...	...	...	...	...	...	0.266 - 0.165

Gudgeon pin

Material ...	...	...	...	...	...	...	EN 32 B
Clearance in small end ins ...	...	...	...	...	...		0.0011 - 0.0005
Clearance in small end mm ...	...	...	...	...	...		0.028 - 0.012
Diameter ins ...	...	...	...	...	...	...	0.6883 - 0.6885
Diameter mm ...	...	...	...	...	...	...	17.4828 - 17.9879
Length ins ...	...	...	...	...	...	...	2.235 - 2.250
Length mm ...	...	...	...	...	...	...	56.75 - 57.15

Torque wrench settings

							Foot-pounds	Kg - m
Alternator rotor nut	...	...	...	...	...	...	50	6.913
Camshaft gear nuts ...	...	...	...	...	...	...	75	10.36
Clutch centre nut ...	...	...	...	...	...	...	60	8.29
Connecting rod nuts (Self locking)	...	...	...	...			18	2.489
Crankcase junction bolts ...	...	...	...	...	...		12	1.659
Crankcase junction stud nuts ...	...	...	...	...			15	2.074
Cylinder barrel nuts ...	...	...	...	...	...		20 - 22	2.76 - 3.04
Cylinder head bolts ...	...	...	...	...	...		18	2.48
Cylinder head stud nuts ...	...	...	...	...	...		18	2.48
Engine sprocket nut ...	...	...	...	...	...		75 - 80	10.36 - 11.06
Fork bottom yoke pinch bolts ...	...	...	...	...			23 - 25	3.2 - 3.45
Front fork cap nuts:								
Steel nut/tapered stanchion ...	...	...	...				80	11.06
Alloy nut/parallel stanchion ...	...	...	...				30	4.14
Fork top yoke pinch bolts ...	...	...	...	...	...		23 - 25	3.2 - 3.45
Front wheel spindle cap bolt ...	...	...	...	...			25	3.456
Kickstart rachet nut ...	...	...	...	...	...		40 - 45	5.5 - 6.2
Main bearing cap nuts (self locking) ...	...	...	...				18	2.489
Oil pressure release valve ...	...	...	...	...	...		35	4.84
Rocker box bolts ...	...	...	...	...	...		6	0.83
Rocker box stud nuts ...	...	...	...	...	...		6	0.83
Rocker spindle domed nuts ...	...	...	...	...			22	3.042
Shock absorber nut ...	...	...	...	...	...		75 - 80	10.36 - 11.06
Sparking plugs ...	...	...	...	...	...		8	1.106
Zener diode nut ...	...	...	...	...	...		2 - 2.3	0.28 - 0.31
Headlamp mounting bolts ...	...	...	...	...	...		10	1.4
Gearbox sprocket nut ...	...	...	...	...	...		58	8.019
Headrace sleeve nut pinch bolt ...	...	...	...	...			15	2.074
Stanchion pinch bolts ...	...	...	...	...	...		25	3.456
Front wheel spindle cap bolts ...	...	...	...	...			25	3.456
Brake cam spindle nuts ...	...	...	...	...	...		20	2.765

1 General description

1 The engines fitted to the Triumph Trident (T150) and the BSA Rocket 3 (A75) are similar in nearly all details. Where differences occur this is pointed out in the text and specifications. It should be noted that with development changes have occurred in both the T150 and the A75 engines but there is little difference between early and late engines. Modifications in the main have been concerned with noise and lubrication. The demonstration engine used for this manual has been removed from a T150 Trident of the late five-speed type.

2 The engine of both models is of unit construction and has a triple bore cylinder block of aluminium. The crankshaft is the 120° type and the crankcase splits into several aluminium alloy mating sections. The engine of the Rocket 3 has a sloping cylinder block, as compared to the vertical cylinder block of the T150 Trident.

3 The gearbox housing is an integral part of the centre portion of the crankcase as is the oil pump.

4 The clutch and primary transmission are housed in separate cases and bolted onto the main crankcase sections.

5 The aluminium alloy cylinder head has cast-in austenitic valve seat inserts. It houses the overhead valves which are operated by rocker arms within detachable alloy rocker boxes bolted to the head. The engines are air cooled.

6 There are six aluminium pushrods to operate the rocker arms. Valve clearances can be adjusted by removing the rocker box covers which provide access to the adjusters on the ends of the rocker arms.

7 The pistons are of die cast aluminium alloy. Two tapered compression rings and one oil scraper ring are fitted to each piston.

8 The connecting rods are of 'H' section R.R.56. Hiduminium alloy with detachable end caps. These have steel backed renewable shell bearings.

9 The inlet and exhaust camshafts are fitted in the upper part of the crankcase and are driven by timing gears from the right-hand side of the crankshaft.

10 The end of the exhaust camshaft operates the adjustable contact breaker and the tachometer drive.

11 The crankshaft is of one piece forged three throw construction. Unlike the "twins" it is set fitted with a flywheel. It is supported by two shell type bearings, a roller bearing at the right-hand side and by a ball bearing on the left-hand side.

12 The aluminium alloy cylinder barrel has Austenitic casting cylinder liners and houses the pressed in aluminium alloy tappet guide blocks.

13 Power from the engine is transmitted through the engine sprocket and triplex primary chain to the clutch sprocket which has an integral shock absorber. From there the power is transmitted via the Borg and Beck diaphragm clutch to the four or five speed constant mesh gearbox and final drive sprocket. Chain tension is controlled by an adjustable chain tensioner immersed within the oil content of the chaincase.

14 The shock absorbers are fitted to smooth out the drive when the engine power is fluctuating under load.

2 Decarbonising

1 Decarbonising is one of the many tasks that can be carried out with the engine in the frame. It also offers the opportunity to examine several parts of the engine for wear.

2 Do not remove the carbon deposits from the engine exhaust ports and combustion chamber until engine performance indicates that this has become necessary. This will only be after a reasonable amount of work has been performed by the machine.

3 Symptoms which indicate a need for decarbonising are as follows:
   A fall off in power and loss of compression
   Noisy operation, pinking, difficult starting
   Increase in petrol consumption
   Engine running hotter than normal

4 To decarbonise proceed as follows:
   Remove the cylinder head
   Remove and service the spark plugs
   Obtain a valve spring compressor

5 Do not use a screwdriver or any steel instrument to remove the carbon on an aluminium surface. A blunt aluminium scraper or a piece of suitably flattened lead solder is a suitable scraping instrument.

6 Leave a ring of carbon around the top of the cylinder bore and when decarbonising the piston crowns leave a ring of carbon around the piston periphery to maintain the seal.

7 Remove the valves and decarbonise the stems, combustion chamber and ports of the cylinder head.

8 All carbon dust should be removed by air jet and the head and valves cleaned in paraffin.

9 Check the valves for pitting and grind-in if necessary.

2.9 Check the valves for pitting

4.1 Lift the engine out from the left hand side

10 Replace items in the reverse order of removal.

3 Operations with engine/gearbox in frame

The Trident and Rocket 3 engines can be worked on quite easily in the frame for a number of operations excluding the obvious, such as work on the crankshaft and main bearings. Operations that can be carried out with the engine in the frame are as follows:

1 Removing and replacing the cylinder head.
2 Removing and replacing the barrels and pistons.
3 Removing and replacing the alternator.
4 Removing and replacing the primary drive components.
5 Removing and replacing the oil pump and points.
6 Removing and replacing the carburettors.
7 Decarbonising the engine.

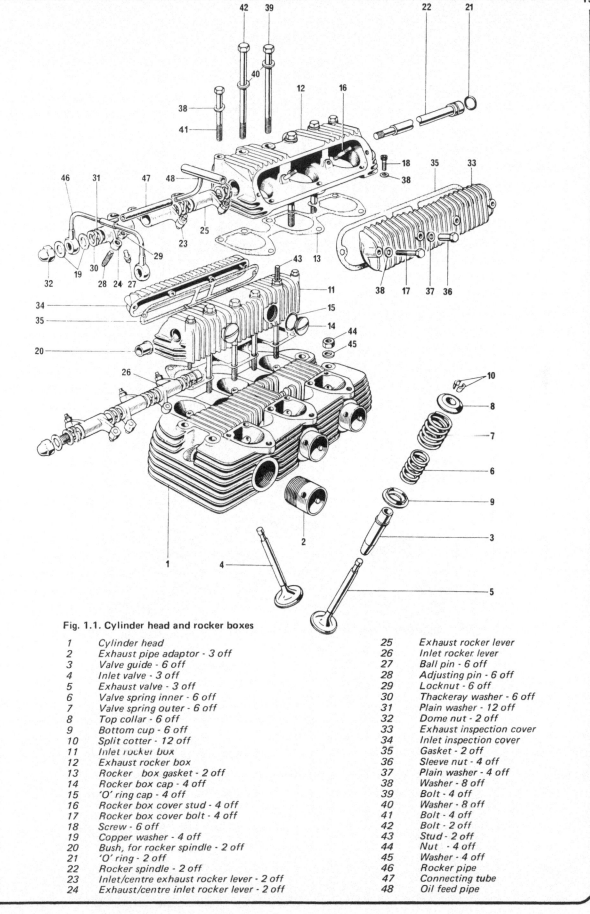

Fig. 1.1. Cylinder head and rocker boxes

1	Cylinder head	25	Exhaust rocker lever
2	Exhaust pipe adaptor - 3 off	26	Inlet rocker lever
3	Valve guide - 6 off	27	Ball pin - 6 off
4	Inlet valve - 3 off	28	Adjusting pin - 6 off
5	Exhaust valve - 3 off	29	Locknut - 6 off
6	Valve spring inner - 6 off	30	Thackeray washer - 6 off
7	Valve spring outer - 6 off	31	Plain washer - 12 off
8	Top collar - 6 off	32	Dome nut - 2 off
9	Bottom cup - 6 off	33	Exhaust inspection cover
10	Split cotter - 12 off	34	Inlet inspection cover
11	Inlet rocker box	35	Gasket - 2 off
12	Exhaust rocker box	36	Sleeve nut - 4 off
13	Rocker box gasket - 2 off	37	Plain washer - 4 off
14	Rocker box cap - 4 off	38	Washer - 8 off
15	'O' ring cap - 4 off	39	Bolt - 4 off
16	Rocker box cover stud - 4 off	40	Washer - 8 off
17	Rocker box cover bolt - 4 off	41	Bolt - 4 off
18	Screw - 6 off	42	Bolt - 2 off
19	Copper washer - 4 off	43	Stud - 2 off
20	Bush, for rocker spindle - 2 off	44	Nut - 4 off
21	'O' ring - 2 off	45	Washer - 4 off
22	Rocker spindle - 2 off	46	Rocker pipe
23	Inlet/centre exhaust rocker lever - 2 off	47	Connecting tube
24	Exhaust/centre inlet rocker lever - 2 off	48	Oil feed pipe

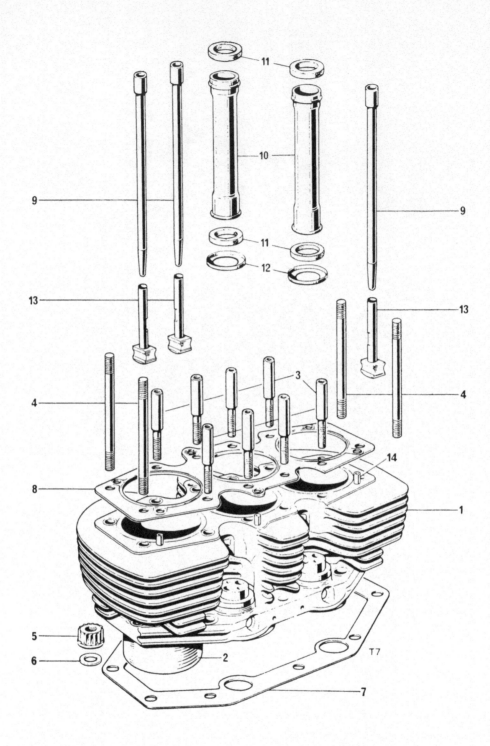

Fig. 1.2. Cylinder barrel, pushrods and tubes, tappet blocks

1	Cylinder barrel c/w liners		8	Cylinder head gasket
2	Cylinder liner - 3 off		9	Pushrod - 6 off
3	Pillar stud - 8 off		10	Pushrod tube - 4 off
4	Stud - 4 off		11	Sealing rubber - 8 off
5	Base nut - 10 off		12	Bottom cup - 4 off
6	Washer - 10 off		13	Tappet - 6 off
7	Cylinder base gasket		14	Dowel - 6 off

4 Method of engine and gearbox removal

1 If a major rebuild is necessary the engine unit will have to be removed to facilitate separating the crankcase for access to the big ends, main bearings, crankshaft etc. This is mostly a one man job until lifting out the engine is required. With an engine weight of approximately 180 lbs the help of a friend is essential whilst actually lifting the unit out of the frame.

2 If lifting equipment is available, this can be attached to the top of the engine using a plate and eye bolt thus making the task so much easier and possibly dispensing with the help of the friend.

3 If the engine is to be lifted out manually, two lifting bars can be located in position on the engine. One fits into the front engine mounting lug and the other in the top rear left-hand side engine plate mounting lug.

4 The lift out now requires two men, one on either side of the crankcase. The engine should be raised and looking at the unit from above it should be turned anticlockwise to avoid the front crankcase lug. Lift the engine out from the left-hand side.

5 Removing the engine/gearbox unit from the frame

1 Before attempting to take the engine/gearbox unit out of the frame it is recommended that the machine should be cleaned with Gunk, Jizer or a similar cleaning agent.

2 Place the machine on the centre stand on firm, level ground. It is expedient to use some strong pieces of timber to stabilise the machine if it is at all unsteady.

3 Unscrew the plastic moulded screw securing the left-hand battery cover and pull the panel off the two locating pegs on the rear frame. Stow any plastic covers in a safe place as they can be easily trodden on during an overhaul and are expensive to replace.

4 Disconnect the fuseholder and remove and stow the fuse. Remove the battery connections and the battery. If the battery is not to be used for a period of time a trickle charge of 1 ampere at weekly intervals should help to keep it in good condition, provided of course that it was in good condition on removal.

5 Remove the petrol tank by following the instructions given in Chapter 2, Section 2.

6 Remove the oil cooler by following the instructions given in Chapter 2, Section 15.

7 Disconnect the rocker feed pipe by disconnecting two domed nuts from the right-hand side of the rocker boxes. The banjo connections can be slid over the rocker spindles. Collect and stow the four copper washers.

8 Remove and stow the right-hand side panel with its associated three Pozidriv screws. The screws should be put into boxes for safe keeping. It is a good policy if new at overhauling machines to put the smaller items in linen or plastic bags with some form of identification. This will make refitting so much easier.

9 Drain the oil from the oil tank using a catchment tank of at least 6 pints (3.41 litres) capacity. Remove the oil cartridge filter end plug from beneath the front of the gearbox outer cover and allow the oil to drain into the catchment tank. If the tank has been filled dispose of the contents before proceeding.

10 Identify the oil pipe stubs which are situated beneath the left-hand side of the centre crankcase section and unscrew the two clips retaining the oil pipes to the stubs. Drain the pipes into the catchment tank.

11 Remove the gearbox drain plug and level plug from beneath the centre crankcase section. The oil can again be drained into the catchment tank. There should be about 1¼ pints (0.710 litres) of oil in the gearbox.

12 Remove the chaincase drain plug which is situated below the centre of the inner chaincase and drain the chaincase into the catchment tank.

13 Disconnect the tachometer cable from its drive box forward of the cylinder barrel.

14 First slacken the left-hand footrest securing nut then remove four Pozidriv screws from the clutch release mechanism inspection cover. Remove the cover and detach the cable from the mechanism. Disconnect the clutch cable by first completely slackening the locknut and cable adjusters to produce sufficient slack to enable the nipple to be removed from the slotted roller.

15 Remove the cylinder head torque stays and lift off the carburettor and air cleaner assemblies as one unit. On some models it is necessary to disconnect the air control cable at the handlebar lever and slip the cable through retaining clips on the frame top rail.

16 The throttle cable should now be removed from the carburettor by loosening off the adjuster and withdrawing the cable from the cable stop. The method of removing the nipple from the throttle actuating lever is to turn it through an angle of 90 degrees and slip the cable and nipple out of the lever by a sideway movement.

17 There are six clips securing the inlet manifold hoses. These should now be slackened off. Remove the rubber buffer from the bottom air filter support lug and the air filter to crankcase breather pipe. Remove the air filter assembly first by removing the connection hoses to facilitate withdrawal from the left side of the machine.

18 Disconnect the three electrical snap connectors for the stator and the three connectors for the contact breaker unit.

19 It is necessary to unscrew the stator wire protector shield to remove it. The top front engine plate nut should be removed from the right-hand side first however.

20 Next remove the exhaust pipes. The first operation is to slacken off the silencer to exhaust pipe clips and then the exhaust pipe to exhaust manifold pinch bolts. It will be necessary to tap the pipes off, preferably using a rubber hammer. The three finned cooling rings should be loosened at the manifold. To remove the manifold, tap it forwards with the rubber hammer.

21 Disconnect the oil pressure switch wire found beneath the crankcase on the right-hand side of the oil pipe stubs.

22 Disconnect the two stop light wires from the left-hand rear engine plate. Remove the rear brake adjusting nut.

23 Remove the final drive chain by uncoupling the split link.

24 Withdraw the oil tank breather pipe from its hollow stub on the chaincase vent chamber inspection cover.

25 It is necessary to disconnect the oil feed pipe from beneath the front of the oil tank to remove the top front engine plate bolt. Next remove the self locking nuts and bolts, plus the large central nut to remove the engine plate.

26 In order to remove the left-hand engine plate the brake pedal must first come off. Remove the nut and spring washer from the brake pedal spindle situated behind the engine plate. The pedal can be removed leaving the actuating rod and lever in situ. Remove the engine plate bolts and lift off the plate complete with footrest. Great care should be taken when removing engine plates and bolts as the engine may shift position.

27 The long engine securing bolt must be removed from under the crankcase. This is accomplished by undoing the left-hand nut and spring washer and removing the bolt from the right-hand side.

28 Locate a spacer between the crankcase lug and the bottom right-hand frame tube. To withdraw this, one nut and spring washer has to be removed from the engine plate to crankcase stud. One bolt and spring washer must be removed and one bolt slackened on the left-hand side detachable engine plate to permit this plate to swing clear of the front engine lug.

29 Section 3 of this Chapter gives the procedures for lifting out the engine; read this before proceeding.

30 Clear a good space for the engine and if a complete stripdown is to follow, it is suggested that paper or rag be spread on the bench.

31 After removal, all the engine bolts should have their nuts spun on lightly and arranged in order to aid reassembly.

5.7 Disconnect the rocker oil feed pipe

5.14 The clutch release mechanism inspection cover

5.15 Remove the cylinder head   torque stays

5.22 Disconnect the two stop light wires

5.23 Remove the chain by uncoupling the split link

5.28 Detachable engine plate

6 Dismantling the engine/gearbox - general

1 The engine should be cleaned, preferably before commencing the stripdown. Care must be taken not to allow any cleaning agent into the exhaust inlets or any exposed parts.
2 Never use undue force to remove any stubborn part. Unless mention is made in the instructions of any special difficulty it may be assumed that if undue force is required the operation is being carried out in the wrong sequence.
3 Dismantling is much easier if an engine stand is constructed. This will allow both hands to be free with the engine held steady on the bench.
4 Whilst dismantling the machine, check for any loose or stray nuts and bolts. There should be a reason for each one of them being loose or off.
5 Have a good look at the cables and connections to see if there is fraying or damage. Apply some electrical adhesive tape at any danger points to give additional protection but always re-place a badly damaged cable with the correct wire. This parti-cularly applies to the HT leads.

7 Dismantling the engine/gearbox - removing the cylinder head, barrel and pistons

1 Remove the high tension leads and spark plugs.
2 Remove the cylinder head bolts. The cylinder head can be lifted squarely off its studs. This can be stored in a rack or recep-tacle, ready for overhaul.
3 The pushrods and tubes can be temporarily left in the cylin-der block.
4 To prevent the tappets falling through into the engine when removing the cylinder block, wrap a piece of insulating tape around each tappet stem. The tappets are loosely located from the underside of their block.
5 To remove the cylinder block, care must be taken to loosen off the ten 3/8 inch diameter nuts in the sequence given in the illustration. Failure to do this could cause distortion to the block.

Stow the nuts and washers in a suitable jar or container and label them.
6 Lift the block upwards and ensure that the pistons coming out of the bores are not damaged by contact with the crankcase mouth. Stuff the crankcase mouth with a clean lint free rag to prevent ingress of stray matter.
7 Identify the location of each tappet and label it accordingly. This will save excessive tappet and cam wear if reinstalled in the same location.
8 Circlip pliers are necessary to remove the gudgeon pin circlips. If the piston is worn the gudgeon pin will come out smoothly but with a fairly new engine hot damp rags will be necessary to heat the piston to enable gudgeon pin removal. Tap out the pin using an alloy drift, taking great care not to damage the pistons or connecting rods. Always support the piston when any pressure is exerted on either the connecting rod or gudgeon pin.
9 Discard the circlips as they should never be re-used. This should eliminate the possibility of their working loose whilst the engine is running, causing the pin to hit the cylinder bore.
10 Mark each piston on the inside to ensure its replacement in its original bore and in the original direction. Failing to install correctly will result in loss of power and high oil consumption.

8 Dismantling the engine/gearbox - removing the contact breaker and timing case

1 The contact breaker assembly is located on the right-hand side of the machine. It is housed in a circular compartment in the outer timing cover and access is gained by removing the chrome cover which is held in position by three screws.
2 The assembly itself comprises the contact breaker back plate on which are mounted three smaller adjustment plates. Three sets of contacts are mounted on the plates.
3 It is advisable before removing the three pillar bolts holding the contact plate to scribe location lines on the plate and housing to assist in reassembly. This will obviate the necessity to re-time the ignition.
4 To remove the plate complete with contacts, undo the three

7.8 Take care to avoid damaging the alloy connecting rods

9.5 An extractor tool is needed at this stage

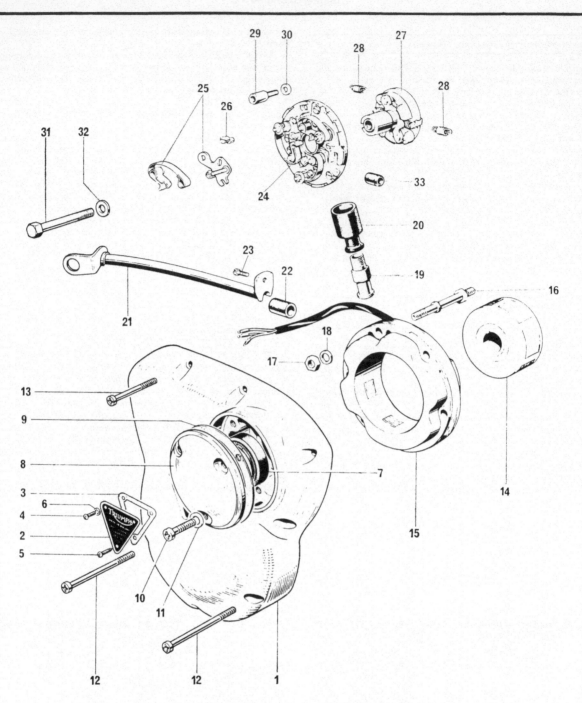

Fig. 1.3. Timing cover and contact breaker (T150 Trident)

1	Timing cover	18	Washer - 3 off
2	Cover	19	Cable seal nut
3	Joint washer	20	Grommet
4	Screw - 2 off	21	Alternator leads cover
5	Pointer screw	22	Rubber ferrule
6	Washer - 3 off	23	Screw
7	Oil seal	24	Contact breaker
8	C.B. cover	25	Contact set - 3 off
9	Cover gasket	26	Adjusting screw - 3 off
10	Screw - 3 off	27	Auto advance unit
11	Washer - 3 off	28	Spring - 2 off
12	Screw - 3 off	29	Pillar bolt - 3 off
13	Screw - 6 off	30	Washer - 3 off
14	Rotor	31	Bolt
15	Stator	32	Washer
16	Alternator mounting stud - 3 off	33	Grommet
17	Alternator mounting nut - 3 off		

pillar bolts. The back plate can hang out of the way suspended on the connecting cables. It is expedient to use a small piece of string or tape to take the weight off the connections.

5   The advance and retard unit is located behind the contact plate. It consists of two bob weights, springs and the contact breaker cam. This assembly is locked into a tapered hole in the exhaust camshaft by its central bolt. Remove the central bolt and give the taper a sharp tap with an alloy drift and the unit will generally fall free. If this is not successful, screw in a long 5/16 UNF bolt and clench it with a mole wrench. Tap the wrench with a hammer to release the unit. BSA service tool for this removal is 60-782, Triumph equivalent D782, 60-0782 or 61-0782.

6   To gain access to the timing gears and generator the timing cover on the right-hand side of the machine must be removed. This is secured by ten screws. Note their positions for refitting as they differ in length. Stow in a box or plastic bag. An impact screwdriver could be necessary for their removal although it is an expensive item to buy. They are worth borrowing for the strip-down if outright purchase is not possible.

7   There may be a small amount of oil trapped in the case so take off the cover over a drip tray.

8   If the point does not break easily a few light taps with a soft headed hammer will help. Pull through the wires from the contact breaker points and remove the timing cover.

9 Dismantling the engine/gearbox - removing the generator and timing pinions

1   Remove the three nuts and washers spaced around the stator. Ease the stator off its studs. The cable sleeve nut is covered by a rubber grommet. Unscrew this nut and slide the cable through.

2   To remove the rotor the locking tab on the large nut on the engine mainshaft must be bent back before attempting to remove the nut.

3   Remove the rotor leaving the key in place. This will prevent the crankshaft timing pinion from turning.

4   Before removing the timing gears take very careful note of the timing marks engraved thereon. If you are in any doubt add a few location marks of your own. Our reassembly illustrations will cover the alignment of these marks in detail.

5   Take out the rotor fixing stud and pull the crankshaft pinion off. You will need an extractor tool at this stage. The service tool recommended is No. 61-6019.

6   The idler gear circlip can now be removed to allow the gear and thrust washer removal.

7   The spindle assembly for the idler gear should now be removed for examination.

8   Leave the gears in position and lock the assembly using a bar of the exact diameter inserted through one of the connecting rods. This will enable the camshaft nuts to be unscrewed. Take care not to damage your crankcase with the bar. The nuts have left-hand threads.

9   Use an extractor to withdraw the crankshaft pinion. Triumph tool 61-6019 is the correct tool for extracting the pinion. Locate the three claws of the tool under the rear face of the pinion and tighten up the large outer collar. The middle thread is the extractor. If no extractor is available the pinions can be removed with great care by using two levers with hooked ends, inserted opposite each other so that they will gradually ease off the pinion. Take care not to damage the gasket faces or the pinion itself.

10   Remove and carefully stow the Woodruff keys.

10 Dismantling the engine/gearbox - removing the primary drive cover, engine sprocket and primary chain

1   The dismantling and reassembly of the primary drive can be

carried out without removing the engine from the frame. This manual assumes that this exercise is being carried out with the engine on its stand and out of the frame.

2   Before removing the primary drive cover ensure the oil has been drained as per Section 4.9 of this Chapter. The oil drain plug is in the lower centre of the inner chaincase.

3   The clutch inspection cover is secured by four screws which should be removed to gain access to the clutch actuating mechanism. If the engine is out of the frame this will already have been removed.

4   Unscrew the locknut and the adjuster from the end of the pullrod.

5   The actuating plate assembly is retained by two anti-vibration spring plates.

6   Completely slacken off the primary chain tensioner and remove the plug to gain access to the lower front countersunk screw (Triumph).

7   It is necessary to remove 14 screws of differing lengths to release the primary drive cover. Note their positions and then bag and stow. You should have 11 Pozidriv, and 3 countersunk.

8   The Triplex primary chain has no connecting links and has to be removed complete with the clutch chain wheel and engine sprocket.

9   Straighten the tag on the lockwasher to the engine sprocket nut and remove nut and washer. The sprocket will require spragging.

10   Unscrew the clutch centre nut. This contains a small oil seal which should not be lost. Withdraw the spacer and thrust bearing.

11   Pull away both the primary chain, the engine sprocket and clutch chain wheel in one unit.

11 Dismantling the engine/gearbox - removing the inner case and clutch

1   The inner case is secured by the following screws or bolts:
   7 'Allan' screws
   2 'Phillips' screws
   2 Bolts
   1 Countersunk screw
   These vary in length so a note should be made of their positions before they are bagged and stowed.

2   Pull off the inner chaincase for clutch removal.

3   Ensure that the large rubber 'O' ring in the oil pump aperature is not damaged or lost.

4   Remove the clutch shaft spacer. The clutch is a sliding fit on the splines of the clutch hub and can be lifted straight off complete with its pullrod.

5   The removal of the oil pump is covered in Chapter 2, Section 11 of this manual.

12 Dismantling the engine/gearbox - removing the small clutch hub, the clutch housing and the final drive sprocket

1   The hub nut requires removal with a box spanner, Tommy bar and hammer. The clutch hub is keyed and tapered to the gearbox mainshaft and calls for the use of extractor D1860 (Triumph) 60-1860 (BSA). The extractor threads into the end of the splined hub. Bag and stow the Woodruff key when the hub is withdrawn. The demonstration model was spragged to allow nut removal.

2   Three countersunk screws require removal in order to release the clutch housing. The breather duct cover is secured to the clutch cover with three additional countersunk screws.

3   To remove the final drive sprocket, first bend back the bent over portion of the tab washer. Place an old chain around the sprocket and fold the chain to make it rigid. The large hexagonal

10.9 The engine sprocket will need   spragging

11.2 Pull off the inner chaincase for clutch removal

12.1 Sprag the pinion to allow nut removal

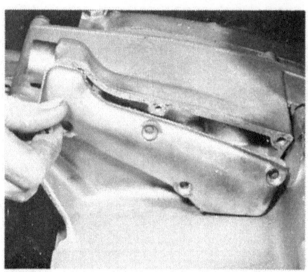

12.2 The breather dust cover is secured to the clutch cover

12.3A First bend back the tab washer

12.3B Place an old chain around the sprocket to lock it

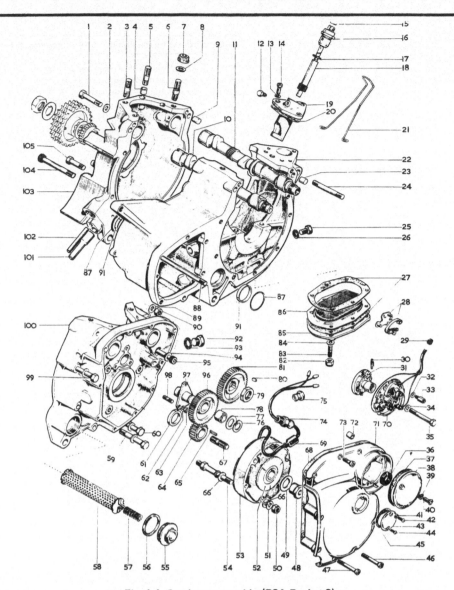

Fig. 1.4. Crankcase assembly (BSA Rocket 3)

1	Bolt - 2 off	28	Contact set	
2	Washer - 8 off	29	Grommet	
3	Stud - 4 off	30	Spring - 2 off	
4	Dowel - 2 off	31	Auto-advance unit	
5	Stud - 2 off	32	Washer - 3 off	
6	Stud - 4 off	33	Pillar bolt - 3 off	
7	Nut - 10 off	34	Washer	
8	Washer - 10 off	35	Bolt	
9	Dowel - 2 off	36	Oil seal	
10	Inlet camshaft	37	Gasket	
11	Exhaust camshaft	38	Cover	
12	Bolt	39	Fibre washer - 3 off	
13	Washer - 2 off	40	Screw - 3 off	
14	Screw - 3 off	41	Washer - 2 off	
15	'O' ring	42	Screw	
16	Housing	43	Timing aperture cover	
17	'O' ring	44	Screw	
18	Driven gear	45	Cover gasket	
19	Housing	46	Screw (short)	
20	Gasket	47	Screw (long) - 4 off	
21	Support wire	48	Rotor nut	
22	Crankcase	49	Lockwasher for rotor nut	
23	Dowel - 2 off	50	Nut - 3 off	
24	Stud - 2 off	51	Washer - 3 off	
25	Oilway plug - 2 off	52	Alternator rotor	
26	Washer - 2 off	53	Alternator stator	
27	Gasket - 2 off			

54	Stator mounting stud - 3 off		- 2 off
55	Main oil filter cap	80	Camshaft pinion key - 2 off
56	Fibre washer	81	Camshaft pinion - 2 off
57	Spring	82	Nut - 6 off
58	Main oil filter	83	Stud - 6 off
59	Crankcase bolt - 4 off	84	Washer - 6 off
60	Crankcase bolt	85	Sump plate
61	Distance piece for timing pinion	86	Filter gauze
62	Washer - 3 off	87	'O' ring
63	Nut - 3 off	88	Stud - 2 off
64	Timing pinion	89	Washer
65	Woodruff key	90	Nut
66	Washer - 3 off	91	'O' ring
67	Stud	92	Timing plug
68	Timing cover gasket	93	Washer
69	Grommet	94	Washer - 7 off
70	Contact breaker assembly	95	Socket screw - 7 off
71	Timing cover	96	Timing pinion
72	Dowel - 2 off	97	Intermediate gear spindle
73	Screw - 5 off	98	Stud - 3 off
74	Cable sleeve nut	99	Bolt
75	Cable clamp	100	Timing side crankcase
76	Circlip	101	Oil return pipe stub
77	Thrust washer	102	Oil feed pipe stub
78	Needle roller bearing	103	Drive-side crankcase
79	Camshaft pinion nut	104	Bolt - 5 off
		105	Bolt

gearbox sprocket nut is 1.66 inches across the flats. If an especially large spanner is not available use spanner No. 61-6061 (Triumph or BSA).

4   With the tab washer and large nut removed the sprocket can be pulled off its splines. 'Hydroseal' is normally used on the splines and the sprocket will have to be removed using extractor No. 61-6046.

13 Dismantling the engine/gearbox - removing and replacing the tachometer drive

1   The tachometer drive is situated at the top of the crankcase centre section in front of the cylinder barrel flange. It is described as a rev-counter drive by BSA.

2   The uncoupling of three Posidriv screws enables the unit to be withdrawn. It consists of a housing, a spindle and a gear to link with another on the exhaust camshaft. Take care not to damage the exhaust face gasket.

3   To further disassemble remove the hexagonal headed locating peg to withdraw the driven gear spindle housing. The spindle can be withdrawn for inspection.

4   There should be no need to replace any part and the only need for stripping is for cleaning and examination. A worn part should be renewed.

5   Points to examine are as follows:
The gear teeth are not damaged
The milled slot engaging with the drive cable thimble is in good condition
The bush in the gear spindle housing is not badly worn. If there is an excessively large amount of sideways movement the bush needs replacing

6   Reassembly is the reverse procedure to disassembly. When refitting the spindle housing ensure that the grooved peg location hole lines up with that in the main body. Refit the pegged screw.

14 Dismantling the engine/gearbox - separating the crankcases

1   The crankcase will divide into three sections. Ensure that all of the following have been removed:
Timing gears (See Chapter 1, Section 9)
Primary drive gear (See Chapter 1, Section 10)
Gear clusters (See Chapter 2, Sections 2 and 3)
Gearbox outer case (See Chapter 2, Sections 2 and 3)
Full flow oil filter (See Chapter 4, Section 12)

2   Removal of the timing side crankcase calls for the removal of the following:
a)   Two bolts at the rear of the crankcase
b)   Two bolts at the front of the crankcase
c)   One bolt at the rear
d)   One bolt at the bottom
e)   One socket headed screw inside the timing case in the top right-hand corner
f)   Two nuts and washers from the studs on either side of the crankcase mouth behind the timing case

3   Removal of the drive side crankcase calls for the removal of the following:
a)   Two bolts at the crankcase mouth
b)   One bolt beneath the case
c)   Five bolts from the front and rear of the case

4   The two outer crankcases are located in dowels and will need tapping off these. Don't lose the dowels. The drive side crankcase section can be tapped off with an aluminium drift and mallet. The timing side, if difficult to remove, can be extracted using a service tool (No. 61-6046).

5   Take hold of the camshafts as the timing side crankcase is removed and stow ready for overhaul.

15 Dismantling the engine/gearbox - removing the crankshaft

1   Earlier models of the Trident and Rocket have two small oil pipes for tappet lubrication on the top of the two main bearing journal caps. These are bent forward to locate in rubber grommets in the front of the crankcase. Removal of these pipes is to extract the securing screws and pull each pipe upwards until it is free from the cap, then turn it away from its cap, push downwards and out of the crankcase. It has been found in practice that this extra oil is not necessary and later models do not have these pipes. If the pipes are carefully blanked off the crankshaft life will be extended.

2   Remove the self-locking nuts and washers from the caps and lift out the complete crankshaft assembly. It may be necessary to gently prise the caps off their studs.

3   Remove the connecting rods from the crankshaft but the rods, bolts and caps must be clearly marked so that they can be replaced in the same position if they are to be reused.

16 Examination and renovation - general

1   Now that the engine is stripped completely, clean all the component parts in a petrol/paraffin mix and examine them carefully for signs of wear or damage. The following Sections will indicate what wear to expect and how to remove and renew the parts concerned, when renewal is necessary.

2   Examine all castings for cracks or other signs of damage. If a crack is found, and it is not possible to obtain a new component, specialist treatment will be necessary to effect a satisfactory repair.

3   Should any studs or internal threads require repair, now is the appropriate time. Care is necessary when withdrawing or replacing studs because the casting may not be too strong at certain points. Beware of overtightening; it is easy to shear a stud by overtightening giving rise to further problems, especially if the stud bottoms.

4   Where internal threads are stripped or badly worn, it is preferable to use a thread insert, rather than tap oversize. Most dealers can provide a thread reclaiming service by the use of Helicoil thread inserts. They enable the original component to be re-used.

17 Main bearings and oil seals - examination and renovation

1   When the bearings have been pressed from their housings, wash them in a petrol/paraffin mix to removal all traces of oil. If there is any play in the ball or roller bearings, or if they do not revolve smoothly, new replacements should be fitted. The bearings should be a tight push fit on the crankshaft assembly and a press fit in the crankcase housings. A proprietary sealant such as Locktite can be used to secure the bearings if there is evidence of a slack fit and yet they are fit for further service.

2   The crankcase oil seal should be renewed as a matter of course, whenever the engine is stripped completely. This will ensure an oiltight engine.

18 Crankshaft assembly - examination and renovation

1   Wash the complete crankshaft assembly with a petrol/paraffin mix to remove all surplus oil. Mark each connecting rod and cap, to ensure they are replaced in exactly the same position, then remove the cap retainer nuts so that the caps and connecting rods can be withdrawn from the crankshaft. Keep the bolts and nuts together in pairs, so that they are replaced in their original order. It is best to unscrew the nuts a turn at a time, to obviate the risk of distortion.

12.4 The sprocket can be pulled off its splines

13.1 The tachometer drive is situated at the top of the crankcase

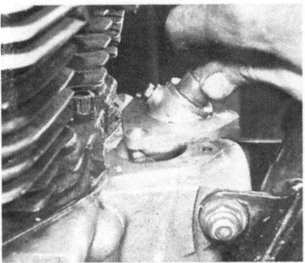

13.2 Removal of the three screws enables the unit to be withdrawn

14.2 Removal of the timing side crankcase

17.1 New replacement bearings should be fitted

18.2 Bearing shells should never be re-used

18.3 If the crankshaft is scored, have it reground

18.4 A reject bearing cap

18.5 No doubt about the need for a replacement here!

19.1 Examine each camshaft

2   Inspect the bearing surfaces for wear. Wear usually takes the form of scuffing or scoring, which will immediately be evident. Bearing shells are cheap to renew; it is wise to renew the shells if there is the slightest question of doubt about the originals.
3   More extensive wear as shown will require specialist attention either by having the crankshaft reground or by fitting a service exchange replacement. If the crankshaft is reground there are four undersize bearing shells available for both big-end and main bearings. They are: 0.010 inches, 0.020 inches, 0.030 inches and 0.040 inches. Regrind sizes for the crankshaft journals are given in the Specifications Section at the beginning of this Chapter.
4   The replaceable white metal shell bearings are prefinished to give the correct diametrical clearance. Under no circumstances should the bearings be scraped or the end cap joint faces filed. The demonstration model cap is a reject.
5   The connecting rod on the demonstration model required renewal as did the other items with illustrations in this Section.

19 Camshafts, tappet followers and timing pinions - examination and renovation

1   Examine each camshaft, checking for wear on the cam form, which is usually evident on the opening flank and on the lobe. If the cams are grooved, or if there are scuff or score marks that cannot be removed by light dressing with an oilstone, the camshaft concerned should be renewed.
2   When extensive wear has necessitated the renewal of a camshaft, the camshaft and tappet followers should be renewed at the same time. It is false economy to use the existing camshaft followers with a new camshaft since they will promote a more

rapid rate of wear.

20 Camshaft and timing pinion bushes - examination and renovation

1   It is unlikely that the camshaft and timing pinion bushes will require attention unless the machine has covered a high mileage. The normal rate of wear is low.
2   The bushes if in need of replacement can be pressed out and in quite normally. The crankcase must always be heated first and well supported. Hot rags in the area of the bush to be replaced can be used to supply the correct heat.

21 Cylinder block - examination and renovation

1   The three cylinder bores should be minutely checked for excessive wear which is normally indicated by a deep ridge around the top of the bore. The cylinder bore will also require attention if there is any deep scoring as this will cause loss of compression and heavy oil consumption.
2   The point of maximum wear in the bores is the top one inch in the direction of rotation. Bore wear at the base of the cylinder is normally slight. If the wear exceeds 0.005 inches (0.127 mm) at the top (rotation) then a rebore with new pistons is essential.
3   Each cylinder bore is fitted with a cast iron liner enabling +0.010, +0.020, +0.030, +0.040 inch rebores to be carried out and for use with oversize pistons and rings. The cast iron liners should stand 0.002 to 0.007 inches proud of the top face of the cylinder barrel.
4   If an engine has been rebored the new size will be marked on

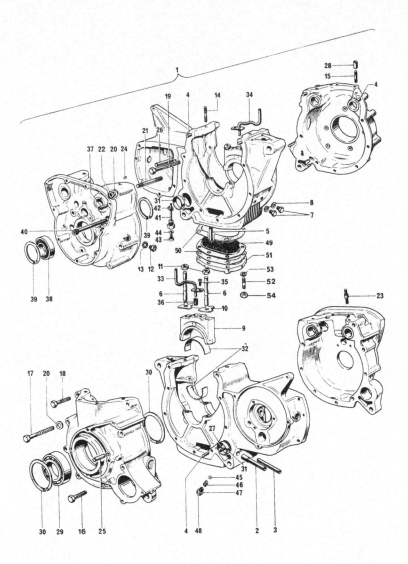

Fig. 1.5. Crankcases (T150 Trident)

1	Crankcases	28	Dowel - 2 off
2	Oil return pipe stub	29	Main bearing
3	Oil feed pipe stub	30	Circlip - 2 off
4	Dowel - 4 off	31	Sleeve at 'O' ring - 2 off
5	Scavenge tube	32	Main bearing shell - 4 off
6	Waisted stud - 4 off	33	Tappet oil feed pipe
7	Main bearing oilways plug - 2 off	34	Tappet oil feed pipe
8	Washer - 2 off	35	Screw - 2 off
9	Main bearing cap	36	Sealing rubber - 4 off
10	Tab washer - 4 off	37	Dowel - 2 off
11	Nut - 4 off	38	Main bearing
12	Timing aperture plug	39	Circlip - 2 off
13	Timing aperture washer	40	Socket screw
14	Cylinder base stud - 4 off	41	Gearbox drain plug
15	Stud - 2 off	42	Fibre washer
16	Clamping bolt - 2 off	43	Level plug
17	Clamping bolt - 9 off	44	Fibre washer
18	Clamping bolt - 2 off	45	Non return ball
19	Clamping bolt	46	Spring
20	Washer - 9 off	47	Joint washer
21	Stud - 2 off	48	Plug
22	Nut - 2 off	49	Scavenge filter
23	Stud - 4 off	50	Gasket - 2 off
24	Peg	51	Cover
25	Spindle	52	Stud - 6 off
26	Selector spindle	53	Spring washer - 6 off
27	Dowel	54	Nut - 6 off

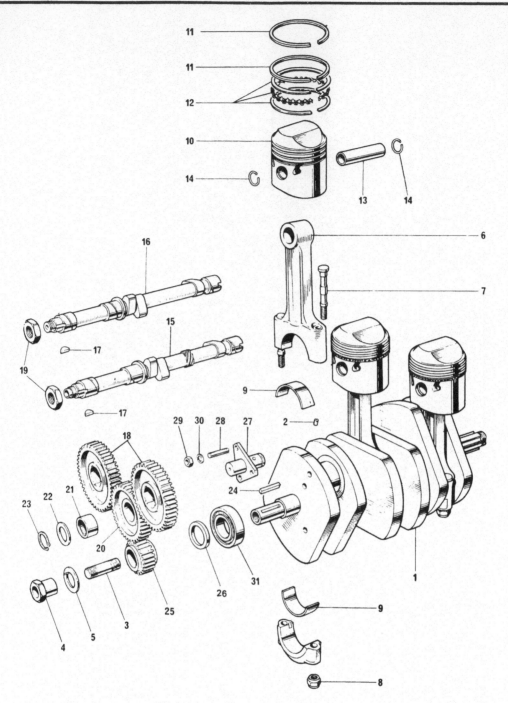

Fig. 1.6. Crankshaft and connecting rods

1	Crankshaft assembly		17	Camshaft key - 2 off
2	Oilway plug - 4 off		18	Camshaft pinion - 2 off
3	Rotor stud		19	Camshaft nut - 2 off
4	Rotor nut		20	Intermediate timing gear
5	Washer		21	Needle roller bearing
6	Con rod assembly - 3 off		22	Thrust washer
7	Con rod bolt - 6 off		23	Circlip
8	Con rod nut - 6 off		24	Timing pinion and rotor key
9	Big end bearing - 6 off		25	Timing pinion
10	Piston assembly - 3 off		26	Distance piece
11	Taper piston ring - 6 off		27	Intermediate gear spindle
12	Scraper ring - 3 off		28	Stud - 3 off
13	Gudgeon pin - 3 off		29	Intermediate gear nut - 3 off
14	Circlip - 6 off		30	Washer - 3 off
15	Exhaust camshaft		31	Main bearing
16	Inlet camshaft			

21.1 The three cylinder bores should be minutely checked

22.2 Clean off each piston crown

22.5 Take care when handling piston rings

23.1 Pass a gudgeon pin through the small end bush

24.2 The valves must be removed with a valve spring compressor

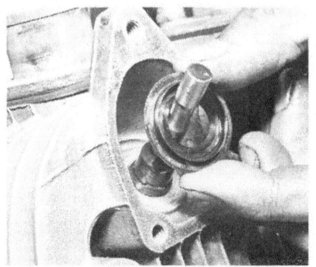
24.3 Take out one valve at a time

top of the piston. If unmarked then it can be assumed to be standard size.

5   Cleaning off the road dirt and painting the fins either matt or pot black will assist heat distribution and will add to the well finished look when the engine is reassembled.

22 Piston and piston rings - examination and renovation

1   The pistons and rings can be overlooked if the cylinder barrel is having a rebore because new pistons and rings will be fitted in the appropriate oversize.

2   If a rebore is not required it does not necessarily follow that the pistons are in good shape. Check for scores, cracks and damage to circlip grooves and renew if necessary. If the pistons are usable clean off each crown and polish them. Carbon does not adhere so readily to a polished surface.

3   Use an old piston ring to scrape carefully around the ring grooves and release any build up of carbon.

4   The outside face of each piston ring should possess a smooth metallic surface and should show no signs of discolouration. If they are discoloured then the ring needs replacing.

5   Take care when handling piston rings as they are brittle and easily broken. They should retain a small amount of spring tension and when placed on a level surface the ring ends should show at least a 0.1875 inch gap.

6   Piston ring gaps can be checked by inserting the ring in the least worn part at the bottom of the cylinder. Use the top of the piston to ensure that it is located squarely. Use a feeler gauge to measure the end gap. When new this should be between 0.009 - 0.013 inches. If the gap is less than 0.009 inches it must be carefully filed to the correct gap.

7   Ring grooves in the piston can be checked with the aid of a new ring. If this is placed end on in the groove and rolled completely around the piston it can be moved in the groove from side to side to test for end play. Excessive end play could mean a new piston.

8   When replacing the rings remember that the compression rings are tapered and it is essential to ensure that the side marked "TOP" is fitted uppermost in each case.

23 Small end bushes - examination and renovation

1   The play in a small end bush can be ascertained by passing the gudgeon pin through the bush - the pin should be a nice sliding fit.

2   If renewal is necessary use the simple draw bolt method as illustrated. Take great care to line up the oil hole in the bush with the hole in the top of the connecting rod.

3   When fitted, it is usually necessary to ream the bush. If the engine is still in the frame put rag around the mouth of the crankcase and ream until the gudgeon pin is a good sliding fit.

24 Cylinder head and valves - dismantling, examination and renovation

1   It is best to remove carbon deposits from the cylinder head before taking out the valves using a piece of soft material to avoid scratching the combustion chambers. Finish with metal polish. It is also advisable to screw in a set of old spark plugs to keep the threads clear.

2   The valves must be removed with a valve spring compressor. The compressor compresses the spring and releases the collets, which are held to the top of the valve stem by a taper in the spring cap.

3   Take out one valve at a time, and before grinding in, check the fit between the valve and the valve guide, which is in the head. Always replace the valves in their correct positions.

4   If the valve guide play is not excessive then regrind the valve. First clean the back of the valve, taking care not to damage the seat. Check each valve for damage to its seat and the seat itself for pocketing which means that the valve is sinking back into the head. If this has occurred get it re-cut at your dealers.

5   Using a small rubber suction pad on a stick and some carborundum paste, apply the sucker to the valve head and a small amount of paste to the valve seat. Spin the grinding tool between the hands and every three or four spins, lift away from the seat; if a little paraffin is put on the valve stem and on the paste it will cut in more easily. The object of this is to re-align the valve and seat and to eradicate any imperfections in it. A grey, dull colour shows when the seating is restored (on both valve and seat).

6   If there is any doubt about the valves, renew them - the large diameter valves are the inlet, the small ones are the exhaust.

7   If the valve guides are badly worn replacements are recommended. Remember that a quite large tolerance can be expected because of the extreme heat the cylinder head experiences.

8   To replace the valve guides, warm the cylinder head in an oven and tap out the guides from the inside. Some guides are bronze; others cast iron. A valve guide drift can be fabricated from a 5 inch length of mild steel bar turned to 5/16 inch diameter for a length of 1 inch. Do not overheat the cylinder head or warpage may occur; seek specialist advice if you are not familiar with this task.

9   The circlips should be removed with the valve guides and refitted when tapping in a new guide. When the circlip is right against the head and in its correct position, a different note in the tapping will occur.

10 If the machine is used for high speeds it is recommended that new valves be fitted as a matter of course. Valves suffer from fatigue more than most parts of the engine.

11 The valve springs should be renewed as a matter of course. They are reasonably cheap and the advantages overcome the cost. Good valve springs are a must for high performance, and make the whole engine run more efficiently.

12 Reassemble the valves in the reverse order of stripping, making sure to oil liberally all moving parts with clean, new oil.

13 Finally make sure that the gasket face of the head is flat, by lapping on an old sheet of glass. (An old car window is perfect, providing it is flat.) Put fine wet and dry rubbing down paper on it and tape down around the edge, then with a rotary motion, rub the head without pressing hard, until a grey matt finish on the jointing surface is achieved.

14 Wash off the surface and dry. Remove the old spark plugs, making sure that no carbon is left in the threads. Wrap the cylinder head in rag and put it aside. The cylinder head is now ready for reassembly.

15 On occasions valve springs have been discovered fitted the wrong way round. This is discovered when valve float occurs at lower than normal revolutions. The close coils of the springs should be fitted towards the head.

25 Cam followers and cam follower guide blocks - examination and renovation

1   The cam followers and guide blocks usually need no attention but a check for wear is necessary in case the machine has been run low in oil, or some similar catastrophy has occurred. Rock the follower in the guide block. It should be a good sliding fit, with very little sideways movement.

2   If excessive wear is found remove the cam follower block and the small location screws on the front and back of the barrel adjacent to the blocks. Use Triumph service tool 61-6008 to drift them out, upwards, taking care not to hit the drift too hard. Cylinder barrels have been known to break by over-zealous hammering during this operation.

3   To replace them use the same drift and after renewing the 'O' ring, grease the block, lightly line up the location hole and drift it in until right home. Repeat this for the other block; refit the locating screws and tighten them.

4   If the cam followers need replacing, they must be replaced in their correct positions, otherwise poor lubrication and damage is likely to occur.

5   Remember that as soon as the engine is started on reassembly all cam and valve gear is running under extreme pressure, so oil all components liberally with clean engine oil.

24.6 If in any doubt about the valves - renew them

24.11 The valve springs should be removed as a matter of course

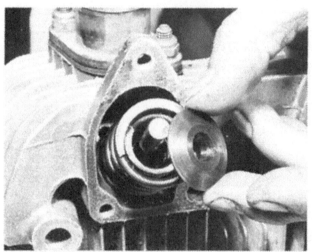

24.12 Reassemble the valve in reverse order of stripping

26.1 Check the pushrods for straightness prior to insertion

26.7 Fit new seals at tops of pushrod tubes

6   When refitting the cam follower stems it is essential that the oil holes in the blocks line up with the hole in the stems to prevent oil starvation.

26 Pushrods, rocker spindles, rocker arms and rocker box - examination and renovation

1   The rocker box assemblies need no attention unless proved to be faulty. They can however be attended to with the engine in position.
2   First remove the engine steady brackets by unscrewing the two nuts securing the brackets to the top of the inlet rocker box and the single nut from the frame lug behind the engine.
3   The rocker boxes are retained in each case by two small end bolts, three nuts from the underside of the cylinder head and by four of the head fixing bolts.
4   Unscrew the two domed nuts securing the rocker oil feed unions to the ends of the rocker spindles and tie the pipe out of the area. Do not lose the copper sealing washers fitted on each side of the unions. If necessary before replacement the washers can be annealed to soften them.
5   Loosen off the valve rocker adjusters completely after removing the rocker box covers.

6 It is necessary to loosen off all of the head fixing bolts in the correct rotation to avoid distorting the cylinder head before removing the four for each rocker box.

7 Remove all of the associated nuts and bolts before lifting out the rocker boxes complete with their spindles and rocker arms.

8 The rocker box spindles are pressed into the rocker box housing. Tap the shaft out from the threaded end leaving the rockers in situ.

9 Make a careful note of the position of the springs and thrust washers before removing the rockers. The spring washers are always fitted next to the shaft pillars not to the rockers. There is no adjustment for end float. This function is performed by the spring washers.

10 The tips of the rocker arms should be examined for wear. If this is evident both the valve clearance adjusters and the ball pins should be removed. The latter should be pressed into place with the drilled flat towards the rocker spindle.

11 The pushrods can have previously been checked for straightness by rolling them on a sheet of glass. If any are bent they should be renewed.

12 Check the end pieces to ensure they are a tight fit on the light alloy tubes. If the end pieces work loose renew the pushrod. It is unlikely that the end pieces will show signs of wear at the point where they make contact with the rocker arms and the tappet followers, unless the machine has covered a very high mileage. Wear will usually take the form of chipping or breaking through the hardening necessitating renewal.

13 Wear in the rocker arm spindles is very rare, unless they have been starved of oil or an extremely high mileage has been covered. To replace, use a soft drift to drive out the spindle and remove the rocker arms and washers, taking note of the location of the various washers.

14 Clean off all the components thoroughly and blow through all the oilways. Remove the oil seals from the spindles and renew them all.

15 It is a good idea, when at this stage, to lap in the cylinder head barrel joint, after just removing the two remaining studs by locking together two nuts to extract them. Lap with a sheet of plate glass and a sheet of very fine wet and dry rubbing down paper, using a rotary motion until a dull, grey consistent finish is achieved over all. Clean thoroughly, then replace the studs.

16 Before commencing assembly of each rocker box, note that there is one plain washer in each box which has a smaller diameter hole. This is the thrust washer through which the smaller diameter of the spindle enters. This is assembled last, against the left-hand inner face of each rocker box.

17 With the rocker box resting base uppermost and using grease to hold the components in place, start from the right and with the spindle just in the box, slide on the first large-holed flat washer, then, moving the spindle in a little at a time, add the other components as per diagram. Repeat this for the other rocker box. Oil liberally.

27 Engine reassembly - general

1 Make sure that every component is clean and that all traces of old gasket have been removed.

2 Make sure that the work area is clean. It is also a good idea to have either brown paper or newspaper to cover the bench.

3 Your tool kit should be clean and the right size for the job. Screwdrivers should have keen, sharp edges. You will also need a hide mallet, a few alloy drifts and a torque wrench. There is nothing worse than having to stop halfway through owing to lack of tools.

4 Finally, you will require new oil seals and gaskets and clean engine oil for the finished engine, also some clean engine oil in an oil can. A well-lit shed, garage, or even kitchen is the other requirement, then you are ready for your engine rebuild.

5 Investigation shows that the greatest amount of wear occurs in the first few seconds after a newly assembled engine has been started up. Factory engineers recommend the use of an oil additive such as S.T.P. or Wynns to be applied to all the bearing surfaces initially. This applies particularly to the camshaft bearings and pistons. Spin the shafts to obtain optimum distribution of the oil.

6 Any oil leaks after reassembly can generally be traced to badly assembled components or unclean mating surfaces.

28 Engine reassembly - reassembling the crankshaft

1 Ensure the crankshaft webs are clear of grease and dirt - this is particularly necessary after a regrinding exercise. On BSA/Triumph crankshafts it is necessary to remove three screwed plugs to allow blowing clear. After blowing clear with an airline secure the plugs in position with Loctite or a suitable alternative.

2 Examine the connecting rod caps and make sure both the front and rear of the bearing shells are scrupulously clean before refitment to the crankshaft journals.

3 Unless regrinding has taken place refit the heads and caps to their original journals and smear all the bearing surfaces with oil. The tab location slots should be fitted adjacent to one another.

4 Refit the bolts and nuts a turn at a time, tightening to the torque figure given in the Specifications Section (Torques) at the beginning of this Chapter.

5 Note the type of nuts fitted to the later models, which are now standardised.

29 Reassembling the crankcases

1 All crankcase sections should be clean and the mating surface clear of old jointing compound.

2 Replace the gearbox high gear and sprocket.

3 Replace the tachometer drive assembly. See Section 14 of this Chapter.

4 Ensure that the two tappet oil feed pipe sealing rubbers have been refitted into the crankcase. They are located at the rear of the tachometer drive and at the cylinder barrel face. This is applicable to earlier models only.

5 Assemble the lower main bearing shells into the crankcase. The locking tab should be at the rear on both shells. The top shell should be fitted into the main bearing caps, ensuring that the locking tabs are to the rear again.

6 Take care not to damage the connecting rods against the crankcase central section when fitting. The large diameter threaded end of the crankshaft goes to the left-hand side.

7 Refit the main bearing caps complete with shells. If they are the ones removed replace them in the same position they originally occupied. Replace the two tab washers and two nuts on each of the caps and tighten to the figure given in the Specifications Section (Torque) at the beginning of this Chapter. Ensure that the crankshaft turns freely. Remember to turn the tabs on the tabwashers over.

8 Older BSA/Triumph machines will have tappet oil feed pipes into the bearing caps but whilst the machine is stripped down carry out the bearing cap modification to remove them. Note the modified cap in the accompanying photograph.

9 Replace the oil pressure release valve and the antidrain valve (see Chapter 4, Section 14). Replace the two oilway blanking plugs and their fibre washers.

10 Fit the sump plate with two new gaskets, one above and one below the gauze filter. The pocketed end of the sump plate is towards the rear of the engine. The removal and replacement of this item is fully covered in Chapter 4, Section 13.

11 Proceed with the assembly in reverse order to the disassembly.

12 The 'O' ring seals fitted in the recesses at either side of the oil filter housing on the central section of the crankcase should be renewed.

13 Lubricate the main bearing and crankshaft supports. Jointing compound should be applied to the mating surface of the left-hand side crankcase section before mating to the central section. Insert the eight securing nuts, bolts and plain washers. Tighten the nuts evenly all around the left-hand crankcase to the torque figures quoted in the Specifications Section at the beginning of this Chapter.

27.5 Lubricate new shells before rebuilding

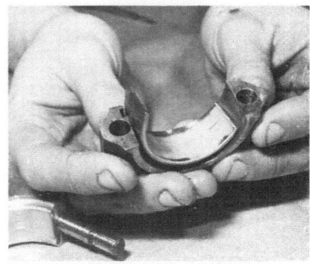

28.3 Refit the rods and caps to their original journals

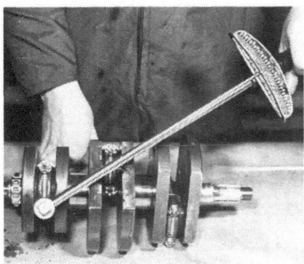

28.4 Tightening to the torque figure given in the Specifications Section

28.5 Note the type of nuts fitted to later models

29.5 Assemble the lower main bearing shells

29.6 Take care not to damage the conrods

29.7A Refit the main bearing caps

29.7B Replace the tab washers

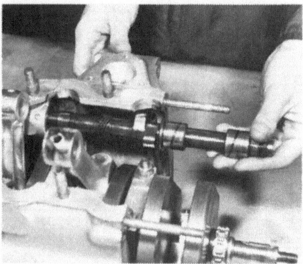
29.16 Fitting the camshafts into position

30.1 Make sure the timing marks coincide exactly

30.3 Engage keys with their original keyways

30.4 Lock connecting rod with a bar when tightening

30.7A Check again that the timing marks coincide

30.7B Circlip and thrust washer can now be replaced on idler

30.8A Carefully knock on the alternator rotor

30.8B Nut has to be torque tightened

30.9A Fit the stator in position, then...

30.9B ... replace the nuts and washers

14 Refit a new 'O' ring in the groove at the opening of the oil pump housing.

15 The right-hand crankcase should now have the mating surface coated with jointing compound.

16 Oil the camshaft journals and roller main bearings before fitting the camshafts into position. Note that the left-end lobes point inwards.

17 Refit the right-hand crankcase using two nuts and plain washers, five bolts and plain washers and a socket head screw. This arrangement of bolts etc., applies to later versions of the Triumph Trident. Differences will be found in the BSA model screws, nuts and bolts etc., but if these items were labelled, bagged and stowed as advised during the stripdown they can be refitted as they came out. Check that both the crankshaft and camshafts rotate quite freely. If they do not then the alignment is at fault and must be rectified.

30 Engine reassembly - replacing the timing pinions, generator and timing cover

1 It is essential to remember when fitting the valve timing pinions to fit them so that the timing marks coincide and also to remember that the cam wheel retaining nuts have left-hand threads. If the disassembly instructions were followed as in Section 9 of this Chapter, then the owner has added his own alignment marks.

2 A service tool assembly D2213 can be used to facilitate the extraction and replacement of both the inlet and exhaust camshaft pinions. It is fitted with an extraction and replacement adapter.

3 When replacing the pinions ensure that the location keys are correctly positioned.

4 To refit, using the adapter tool assembly, screw the adaptor into the assembly bolt and then onto the camshaft. Lubricate the pinion before assembly and screw the adapter body onto it (the thread is left-hand). Slide the pinion and body over the replacer. Align the key to the keyway opposite the timing mark before screwing on the replacer nut and washer. The demonstration model nuts were fitted without the service tool. The connecting rod was held with a bar and a length of timber.

5 Refitting of the crankshaft pinion can be aided by the use of service tool 61-6024. This is a tubular drift and guide.

6 Refit the spacer and refit the key to the shaft. Fit the pinion with the chamfer and timing dot outwards. First screw the guide onto the crankshaft sliding the lubricated pinion over it. Align the key and keyway and drive the pinion onto the crankshaft.

7 Ensure the lubrication of the intermediate spindle and assemble the intermediate wheel so that the timing marks on the three wheels now coincide. The timing marks will only coincide every 94th revolution. The intermediate pinion is secured by a circlip and thrust washer which can now be replaced.

8 Refit the alternator in the reverse procedure to that given in Section 9 of this Chapter. Ensure the locating key is in position before refitting the rotor. Fit a new tab washer and then the large nut on the engine mainshaft. This requires to be torque tightened to 50 ft lbs (6.9 kg m). Turn over the tab on the tab washer after tightening. Apply a smear of Loctite grade AV to the rotor fixing stud before refitting.

9 Fit the stator in position over its three studs which are spaced around the rotor. It should be eased in carefully to prevent damage. Fit the nuts and washers securing it to the studs. Take care to refit the connecting cable correctly. The cable sleeve nut is covered by a rubber grommet. The cable must be fed through and the nut retightened. Loctite grade AV should also be applied to the three stator fixing nuts.

10 The air gap between the rotor and stator should be equal all round. Check with a 0.010 inch feeler gauge. It is essential to correct any misalignment.

11 Ensure the timing cover is clean and free from old cement on crankcase mating surfaces and apply a thin coating of 'Loctite Plastic Gasket' to these surfaces. Refit the 10 timing cover screws. These screws vary in length and in some models range in type. Tighten them in an even sequence to avoid distorting the cover.

12 Refit the advance and retard mechanism and then the contact breaker assembly to the instructions given in Chapter 5, Section 4. Refitting the auto advance unit on its taper will again require the use of service tool 60-782 or its equivalent.

13 The ignition will have to be reset as per Chapter 5, Section 8.

31 Engine reassembly - reassembling the primary drive

1 Information on the stripdown and rebuild of the clutch mechanism is given in Chapter 3, Sections 2 and 3 of this manual.

2 Insert the clutch pull rod before mounting the clutch on the machine.

3 Fit a new oil seal into the clutch case with the open side facing the reassembled clutch.

4 Insert the small splined clutch hub into the oil seal without damaging the seal.

5 Slide the clutch hub with the clutch case onto the gearbox mainshaft. Apply Loctite onto the thread and tighten the nut to 60 ft lbs (8.29 kg m).

6 On early models (BSA) the outer clutch oil seal was fitted with the open side away from the clutch and this necessitated fitting the clutch oil seal sleeve onto the shaft before fitting the inner cover. With later models the oil seal is fitted with the open side to the clutch and the sleeve can be fitted after the case. The sleeve should be glued to the clutch (BSA) with Permatex 300 or equivalent to eliminate any possibility of its working loose, allowing oil to seep into the clutch case.

7 On both clean surfaces of the chaincase apply jointing compound and then a new gasket. Fit the inner cover and tighten the screws evenly to secure.

8 The shock absorber rubbers should be fitted to instructions given in Chapter 3, Section 4. Fit either six hexagon headed bolts with three tab washers or six countersunk headed screws as on earlier BSA models.

9 Refit the oil pump using a new grease-smeared gasket and tighten the four fixing screws. The longest one is through the dowel location. Look out for the 'O' rings when refitting the pump. Replace the idler gear and crankshaft pinion after lightly oiling the idler gear spindle.

10 On reassembly of the chain wheel and engine sprocket the engine sprocket fits with its boss towards the main bearing.

11 Offer the chain wheel, engine sprocket and primary chain as a set, aligning the splines so that the set can be slid home. Tap the sprockets into position with a soft mallet and fit the tab washer and nut at the engine sprocket. The tab on the nut should be tapped over when the nut has been tightened. Ensure all of the thrust washers are fitted in their correct sequence.

12 Fit the plain steel washer at the chain wheen centre, put the oil seal protector over the clutch pull rod threads and fit the securing nut. This should be fully tightened.

13 Fit the lightly oiled face needle roller bearing and large plain steel washer.

14 The primary chain is not adjustable. Wear in the primary chain can be taken up by means of a rubber faced slipper blade below the bottom ring of the chain.

15 Free movement in the chain can be felt with a finger after removing the top inspection plug and pushing down on the kickstarter to apply pressure to the tensioner blades. Do not move the chain by using the starter pedal whilst one's finger is inside the case. The correct chain adjustment is 3/8 inches (0.9 cm) free movement. To reduce the amount of slack remove the plug with the extended head from the front bottom of the case and tighten the slotted adjuster nut at the front end of the tensioner with a screwdriver. Replace the plug with care to avoid damage to the 'O' ring.

16 Before running the engine remember to refill with 5/8 pint (350 cc) of engine oil to bring the level up to the point of efficient chain lubrication. The engine will breath through the main bearing when the engine turns over to maintain the correct level.

30.9C The cable must be fed through

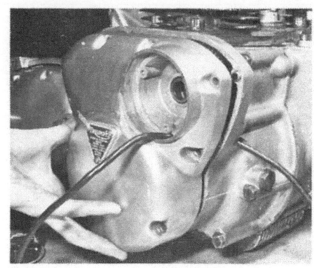

30.11 Ensure the timing cover joint is clean

31.11 Offer the chainwheel etc., as a set

31.12 Fit the securing nut

31.15 Remove the plug from the front

33.2 Use new gudgeon pin circlips

32 Engine reassembly - replacing the outer primary chaincase cover

1   Assembly and readjustment of the clutch operating mechanism is given in Chapter 3, Section 3 of this manual.
2   Prepare the primary chaincase for refitting by removing all remnants of cement and checking that the mating faces are free and level.
3   Ensure that the two locating dowels are fitted and then position the gasket.
4   Fit the outer cover carefully over the oil pump idler spindle, the clutch pull rod and the two dowels.
5   The two middle Posidriv screws can be fitted first followed by nine more Posidriv screws and three countersunk screws. Tighten the screws in diagonal sequence to avoid distortion of the chaincase.
6   The left-hand footrest can now be refitted followed by a large washer and nut
7   Replace the left-hand exhaust pipe and tighten the manifold clamp bolt and then the silencer clamp bolt.
8   The clutch cover can be replaced after clutch adjustment. Ensure that the oil seal in the centre of the cover is not damaged. Ensure the surfaces are clean and tap the clutch cover gently into position until the spigot locates. Fit and tighten the three retaining screws. Fitting the clutch cover at this stage will ensure that no oil, grease or foreign bodies can gain access to the clutch area. The screws can be loosely fitted and removed when the engine is back in the frame and the clutch cable is to be refitted.

33 Engine reassembly - replacing the pistons and cylinder block

1   Each component should be well lubricated before assembly. An additive such as Wynns or STP will help with the camshaft bearings and pistons. In no case should dirt or foreign bodies be permitted to fall into the reassembled sections. A new base gasket is required for the barrel section.
2   Replace the pistons on the connecting rods so that they are the same way round as previously. Use new gudgeon pin circlips and press them well into the grooves.
3   The tappets should be reassembled into the tappet guide blocks in the position from which they were removed. To prevent their falling through again use tape or an elastic band to secure them in position. Oil holes in the tappet stems should line up with the holes in the blocks.
4   Piston slippers, part number 61-6031 or alternative, can be used to compress the piston rings to a point where they are just free to move.
5   Gently lower the cylinder block into position. The slippers are displaced as the pistons enter the bore and can be removed. With the barrel on the crankcase fit and tighten the 10 fixing nuts a little at a time until a torque figure of 20 - 22 ft lbs is achieved (2.76 - 3.04 kg m).
6   On the demonstration model it was discovered that tightening the cylinder barrel nuts was the job for a thinner than normal spanner. The spanner eventually used was a ½ inch thin A.F. ring spanner.

34 Engine reassembly - replacing the cylinder head, push rod tubes and rocker boxes

1   Replace a new cylinder head gasket on the cylinder head studs with the ribs downwards and lower the head squarely over the studs. Refit the four outer stud nuts loosely. The valves and springs should previously have been replaced, care having been taken to correctly fit the tapered collets.
2   Insert the push rod tubes and tube seals (new ones) into position, then the push rods, the single ones being on the drive side. Line up the push rods evenly. The rods should be reinserted in their original positions by following the markings made at the dismantling stage.

33.2A Replace the pistons on the connecting rods

33.2B Each component should be well oiled before assembly

33.3 Use tape or an elastic band to secure them in position

33.5 Gently lower the cylinder barrel into position

34.1 Replace a new cylinder head gasket

34.2A Insert the pushrod tubes

34.2B Line up the pushrods evenly

34.3 The rocker boxes should be fitted with new gaskets

34.4 Cylinder head bolts are torque tightened in correct sequence

3   The rocker boxes should be fitted with new gaskets coated with gasket cement on one side only before replacement. Take care that the push rod tubes are correctly positioned, removing the circular inspection caps so that the upper ends of the push rods can be fitted under their corresponding rocker arms. The push rods must engage correctly with the ends of the tappet followers.

4   Fit the remaining eight cylinder head bolts through the rocker boxes and tighten all the bolts in correct sequence. Cylinder head bolts are torque tightened to 18 ft lbs (2.48 kg m). Cylinder head stud nuts are tightened to the same figures. The longer bolts are fitted on the outside. The rear ones are special bolts to take the engine steady stays.

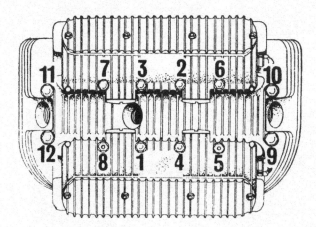

Fig. 1.7. Tightening sequence for cylinder head bolts

5   Tighten the rocker box end bolts and the six Allen screws on the inside of the rocker boxes. Replace the push rod inspection covers. The valve clearances will be reset to Section 38 of this Chapter. The fitting of the inspection covers will keep foreign bodies from falling into the area whilst refitting the engine. They can be removed at a later stage.

6   Care should be taken when fitting head gaskets made from Clingite.

7   Small strands of wire can fall into the combustion chamber and cause pre-ignition. Before fitting these gaskets remove any loose strands of wire no matter how small from the gasket. Do this on the bench not the engine.

35 Engine reassembly - adjusting the valve rocker clearances

1   Adjustment of the valve rocker clearances should only be made with a cold engine. The adjustment can be carried out without other engine overhaul.

2   Check the clearances and readjust if necessary every 3 - 4,000 miles.

3   Remove the inlet and exhaust rocker inspection covers. They are retained by four nuts and bolts.

4   The adjustment tools are carried in the toolkit. A ring spanner and a tappet spanner are needed and also a feeler gauge. The rocker clearances are: Inlet 0.006 inches, Exhaust 0.008 inches.

5   Remove the H.T. leads and spark plugs if they have previously been refitted. Select top gear and turn the rear wheel for crankshaft positioning.

6   Commencing with the inlet camshaft turn the engine until any two rocker arms are 'on the rock'. This means that the two associated valves are open by equal amounts (which is approximately 0.0625 inches). One valve is almost closed and the other just opening. The third valve is the one for resetting. This should be on the base circle. Set the clearances on the inlet side first.

35.7 Tighten the locknut and recheck the gap

35.9 Use new gaskets at cover joint

35.9A Replace the inspection covers

7   Proceed to adjust as follows:
Slacken the locknut
Insert the feeler gauge of the correct thickness between the rocker arm and the valve stem and rotate the adjuster either way to obtain the correct gap. The gauge should be tight but slide freely. Tighten the locknut and recheck the gap.

8   Turn the engine over each time until all the valves are positioned and tightened.

9   Replace gaskets and inspection covers. Refit the spark plugs and H.T. connections.

10 Three different types of tappet adjusters can be found on Triumph Trident machines. Earlier models used a 0.25 inch radius type to be followed by one with a 0.375 inch radius.

11 Latest models use a floating ball bearing system (Ducati type) This also applies to the BSA.

12 The latest system greatly extends the life of the valves. If you approach your dealer for a modification kit, the later valves must go with the adjusters.

13 With the floating ball system it is more difficult to carry out tappet resetting. Noisy tappets will require a complete resetting up exercise.

36 Replacing the engine in the frame

1   The task of replacing the Triumph/BSA engine unit in the frame is the reversal of removal. The first method of replacement is for the BSA Rocket 3.

2   The BSA engine is inclined in the frame at a different angle from the Triumph unit. In either case a lifting attachment is recommended as the engine weighs 180 lbs. Your main dealer will probably have one on a loan/deposit basis but otherwise you will need the assistance of a male friend.

3   Lift the engine into the frame from the left-hand side, the front being lifted slightly higher than the back to give clearance to the front engine lug over the frame lug.

4   When refitting the mounting bolts it may be necessary to juggle the engine about to position it to receive the bolts. Do not omit the packing pieces which fit on the left-hand side of the front engine mounting lug, the bottom lug and the left-hand rear mounting plate complete with footrests. The spacers fit between the mounting plate and the swinging arm plate. Make sure the spacers are in their correct positions.

5   Replace all nuts and washers and correctly tighten them.

6   The second method of replacement is for the Triumph Trident.

7   This can be lifted into the frame with two extremely strong lifting bars which can be located one in the front engine mounting lug and one in the top side engine mounting lug. The operation of lifting in requires two people, one on either side of the crankcase.

8   Lift the unit in gearbox first from the left side. The front of the unit can be swung round into position.

9   Replace the bottom mounting stud, from the right-hand side ensuring that the spacer is fitted in the correct position between the crankcase lug and bottom frame tube on the right-hand side. Replace the spring washer and nut.

10 Line the front detachable engine plate up with the crankcase lug and replace the stud, spring washer and nut. Replace the bolt and spring washer and tighten the remaining bolt.

11 Replace the left-hand engine plate ensuring that the two spacers are refitted between the rear crankcase lugs and the engine plate. Refit the remaining bolts, washers and self locking nuts and the large swinging arm lug bolt and washer.

12 Replace the right-hand engine plate and secure this with its bolts, washers and self locking nuts and the large central collar nut.

13 The nut at the top of the front engine plate bolt need not be replaced at this stage.

37 Final reconnections and adjustments

1   With the engine refitted to the frame replace the rear brake

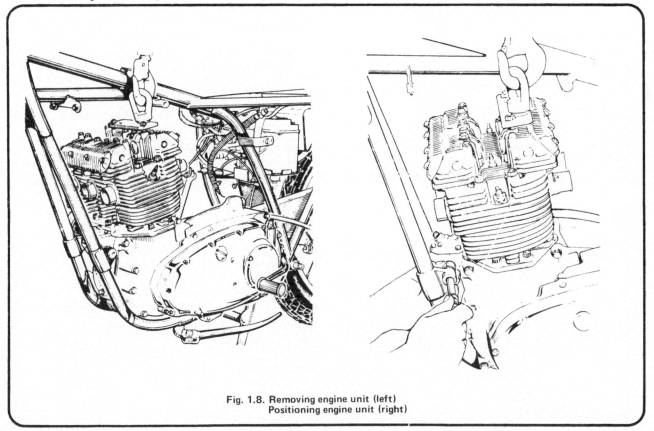

Fig. 1.8. Removing engine unit (left)
Positioning engine unit (right)

pedal. Position the rear brake actuating lever over the squared end of the brake pedal spindle, then replace the nut and spring washer.

2   Refit and retighten the left-hand footrest.

3   Slide the air filter to clutch cover rubber pipe over the crankcase sleeve.

4   Connect both the stator and contact breaker leads (colour to colour). Position the stator lead shield over the top front right side engine plate bolt and refit the self locking nut and slot headed screw.

5   Refit the rear chain over the gearbox sprocket and over the rear wheel sprocket. Fit the split link ensuring that the closed end of the link is to the front of the machine when positioned on the top run of the chain.

6   Reconnect the oil pipes beneath the engine and tighten the securing clips. Check that the oil feed pipe from the bottom of the oil tank leads to the top small stud below the crankcase. The oil pipe from the larger bottom stub connects to the rocker feed pipe. The pipe from the front of the rocker feed connection leads to the left-hand side of the oil cooler and the right-hand side pipe from the oil cooler connects to the return union at the top front of the oil tank.

7   Ensure that the high tension connections from spark plugs to coils are connected correctly. The correct sequence of connections is as follows:

Black/Yellow C.B. lead connects from left-hand cylinder to the top coil when viewing from the right side of the machine.

Red/Black C.B. lead connects from the centre cylinder to the bottom right coil.

Black/White C.B. lead connects from the right-hand cylinder to the bottom left coil.

8   Refill the oil tank and gearbox with the correct amount and grade of oil as given in the Specifications Section. Before the initial start add one half pint of engine oil to the sump. This can be poured through the timing plug aperture in the right-hand crankcase.

9   Replace the exhaust system to the reverse procedure of the removal as given in Chapter 4, Section 9. Make doubly sure that the finned clips have been well and truly tightened to obviate any physical damage.

10   Replace the carburettors and synchronise to the procedure given in Chapter 4, Section 5, 6 and 7.

11   Ensure that the oil warning switch lead is reconnected onto the switch positioned beneath the crankcase, adjacent to the oil pipe unions. Fit the engine shield (BSA) by hooking the back of the shield over the frame cross tube and swinging it up until the slotted bracket fits behind the nuts and washers of the bottom engine bolt. This bottom engine bolt should now be tightened to secure the plate plus the engine bottom lug.

12   The clutch cable should be fed through the frame clips and through the adjuster in the primary chaincase to the lever toggles. Adjust to the limits given in Chapter 3, Section 3.

38 Starting and running the rebuilt engine

1   Switch on the ignition and run the engine slowly for the first few minutes, especially if it has been rebored. Remove the cap from the top of the oil tank and check that oil is returning. There may be some initial delay whilst the pressure builds up and oil circulates throughout the system, but if none appears after the first few minutes running, stop the engine and investigate the cause. If the pressure release valve is unscrewed a few threads, oil should ooze from the joint if the oil pump is building up pressure.

2   Check that all controls function correctly and that the generator is indicating a charge on the ammeter. Check for any oil leaks or blowing gaskets.

3   Before taking the machine on the road, check that all the legal requirements are fulfilled and that items such as the horn, speedometer and lighting equipment are in full working order. Remember that if a number of new parts have been fitted, some running-in will be necessary. If the overhaul has included a rebore, the running-in period must be extended to at least 500 miles, making maximum use of the gearbox so that the engine runs on a light load. Speeds can be worked up gradually until full performance is obtainable by the time the running-in period is completed.

4   Do not tamper with the exhaust system under the mistaken belief that removal of the baffles or replacement with a quite different type of silencer will give a significant gain in performance. Although a changed exhaust note may give the illusion of greater speed, in a great many cases quite the reverse occurs in practice. It is therefore best to adhere to the manufacturer's specification.

5   If using the machine at night on the road please wear something bright so that the rider can be seen. The main dealer will stock some very colourful riding gear that could save a rider's life.

39 Engine modifications and tuning

1   The Triumph and BSA engines fitted to the models covered by this manual can be tuned to give even higher performance and yet retain their very high standard of mechanical reliability.

2   Many special parts are available, both from the manufacturer and from a number of specialists, for boosting engine performance. The parts include high compression pistons, high lift camshafts and transistorised ignition systems.

3   Publications are available to provide information on the ways the engines can be modified to give increased power output.

4   The average machine owner is advised however that a certain amount of skill, equipment and experience is necessary if an engine is to be developed in this manner and still retain the original standard of mechanical reliability. Often it is advisable to entrust this type of work to an acknowledged specialist and therefore obtain the benefit of his experience.

40 Fault diagnosis: engine

Symptom	Cause	Remedy
Engine will not turn over	Clutch slip	Check and adjust clutch.
	Mechanical damage	Check whether valves are operating correctly and dismantle if necessary.
	Discharged battery (T160 model only)	Remove and charge battery.
Engine turns over but will not start	No spark at plugs	Remove plugs and check. Check whether battery is discharged.
	No fuel reaching engine	Check fuel system.
	Too much fuel reaching engine	Check fuel system. Remove plugs and turn engine over several times before replacing.
Engine fires but runs unevenly	Ignition and/or fuel system fault	Check systems as though engine will not start.
	Incorrect valve clearances	Check and reset.
	Burnt or sticking valves	Check for loss of compression.
	Blowing cylinder head gasket	See above.
Lack of power	Incorrect ignition timing	Check accuracy of setting.
	Valve timing not correct	Check timing mark alignment on timing pinions.
	Badly worn cylinder barrel and pistons	Fit new rings and pistons after rebore.
High oil consumption	Oil leaks from engine/gear unit	Trace source of leak and rectify.
	Worn cylinder bores	See above.
	Worn valve guides	Replace guides.
Excessive mechanical noise	Failure of lubrication system	Stop engine and do not run until fault located and rectified.
	Incorrect valve clearances	Check and re-adjust.
	Worn cylinder barrel (piston slap)	Rebore and fit oversize pistons.
	Worn big end bearings (knock)	Fit new bearing shells.
	Worn main bearings (rumble)	Fit new journal bearings.

Chapter 2 Gearbox

Contents

Specifications

4 speed gearboxes

Gearbox

Internal ratios (std)

4th (top)	1.00 : 1
3rd	1.19 : 1
2nd	1.69 : 1
1st (bottom)	2.44 : 1

Overall ratios (4 speed gearbox)

4th (top)	4.89 ; 4.98 (71)
3rd	5.83 : 5.95 (71)
2nd	8.3 ; 8.42 (71)
1st (bottom	11.95 ; 12.15 (71)

Overall ratios (5 speed gearbox)

5th	4.98
4th	5.93
3rd	6.97
2nd	12.9

Gear details

Mainshaft high gear

Bore diameter (bush fitted) inches	0.8135 - 0.8145
Bore diameter (bush fitted) mm	20.6629 - 20.6883
Working clearance on shaft inches	0.0032 - 0.0047
Working clearance on shaft mm	0.08128 - 0.1194
Bush length inches	2.25
Bush length mm	57.15

Layshaft low gear

Bore diameter (bush fitted) inches	0.8135 - 0.8145
Bore diameter (bush fitted) mm	20.6629 - 20.6883
Working clearance on shaft inches	0.0025 - 0.0045
Working clearance on shaft mm	0.0635 - 0.127

Gearbox shafts

Mainshaft

Left end diameter inches	0.8098 - 0.8103
Left end diameter mm	20.5689 - 20.5816
Right end diameter inches	0.7494 - 0.7498
Right end diameter mm	19.0348 - 19.044
Length inches	10 21/64

Length mm	262.3337
Layshaft	
Left end diameter inches	0.6845 - 0.6850
Left end diameter mm	17.4063 - 17.419
Right end diameter inches	0.6845 - 0.6850
Right end diameter mm	17.4063 - 17.419
Length inches	6 41/64
Length mm	168.6941
Camplate plunger spring	
Free length inches	2 21/32
Free length mm	67.4675
Number of working coils	27
Spring rate lbs/inches	9
Spring rate kg/sq cm	0.633
Working range lbs	7.5 - 11.5
Working range kgm	3.405 - 5.221

Bearings

High gear bearing (ball journal) inches	1¼ x 2½ x 5/8
High gear bearing (ball journal) mm	31.75 x 63.4 x 15.875
Mainshaft bearing (ball journal) inches	¾ x 1 7/8 x 9/16
Mainshaft bearing (ball journal) mm	19.05 x 47.625 x 14.282
Layshaft bearing (left) (needle roller) inches	11/16 x 7/8 x ¾
Layshaft bearing (left) (needle roller) mm	17.46 x 22.27 x 19.05
Layshaft bearing (right) (needle roller) inches	11/16 x 7/8 x ¾
Layshaft bearing (right) (needle roller) mm	17.46 x 22.27 x 19.05

Kickstart operating mechanism

Bush bore diameter inches	0.751 - 0.752
Bush bore diameter inches	19.0754 - 19.1008
Spindle working clearance in bush inches	0.003 - 0.005
Spindle working clearance in bush mm	0.0762 - 0.127
Ratchet spring free length inches	0.5
Ratchet spring free length mm	12.7

Gearchange mechanism

Plunger springs	
Number of working coils	12
Free length inches	1.25
Free length mm	31.75
Inner bush bore diameter inches	0.6245 - 06255
Inner bush bore diameter mm	15.7423 - 15.8877
Clearance on shaft inches	0.0007 - 0.0032
Clearance on shaft mm	0.01778 - 0.08128
Outer bush bore diameter inches	0.0005 - 0.0025
Outer bush bore diameter mm	0.0127 - 0.0635
Clearance on shaft inches	0.005 - 0.0025
Clearance on shaft mm	0.0127 - 0.0635
Quadrant return springs number of working coils	9½
Free length inches	1.75
Free length mm	44.45

Chains

Primary 3/8 ins pitch*, triplex links	82
Secondary 5/8 ins pitch x 3/8 inch wide links	104

*number may vary sligthly according to machine specification

1 Gearbox - general description

1   The gearbox can be dismantled without interfering with the timing gear housing. If however it is necessary to remove the gear clusters then first remove the clutch (see Chapter 1, Section 10). The kickstart assembly can be serviced with the engine in situ.

2   The demonstration model used in the illustrations for this manual was fitted with a five speed assembly. Earlier versions of the Trident and Rocket 3 were fitted with a four speed gearbox. However, apart from the addition of the extra gear pinions the four and five speed gearboxes are similar in most respects.

3   The gearbox is easy to work on and will normally require attention only after a very high mileage has been covered or if the oil level has not been maintained.

4   To meet the demands of high performance such as in road racing, close ratio gear clusters are available which with little or no modification will replace the standard gears. For details of such a change consult a dealer who has the appropriate Triumph or BSA replacement parts lists.

5   The Trident gearbox inner and outer covers are of aluminium alloy D.T.D. 424 to give strength and rigidity. The gears are high quality nickel steel which has been specially case hardened.

6   At each end of the mainshaft are heavy duty ball races. The layshaft has special needle roller bearings which are pressed into the casing and inner cover.

7   The single plate diaphragm clutch has a splined hub which is keyed to the left end of the gearbox mainshaft.

8   The quadrants for the kickstart and gearchange are housed in the gearbox outer cover.

9   The gearbox is operated by a pedal on the right side of the machine. The pedal is spline fitted to the gearchange spindle and the plunger housing. Two chamfered plungers with springs fit into the housing in such a way that as the gear pedal is operated, the plungers locate in the teeth at the outboard end of the quadrant. The pivotted quadrant mates with the captive pinion of the camplate.

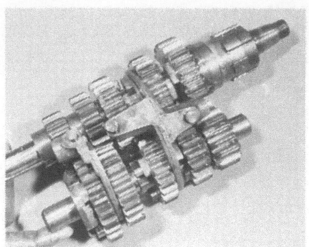

1.2 The five-speed gear assembly

10 Four sliding pinions are moved along the mainshaft and layshaft by the selector forks as the forks are moved together and apart against the track of a camplate.

11 When the pedal is depressed to engage first gear the camplate is turned clockwise moving the layshaft selector fork to mesh the layshaft sliding gear with the layshaft first gear. When the second gear is selected by lifting the pedal the camplate is moved this time anticlockwise to move the layshaft sliding gear to a neutral position and the mainshaft sliding gear into mesh with mainshaft third gear. Similar engagements occur for the other gear selections.

12 Throughout the range of gear pedal movements, the gear pedal spindle and plunger housing return to the original position ready for the next selection. This provides what is known as the positive stop action, so that gears can be selected only in sequence.

2 Dismantling the engine/gearbox - removing the gearbox outer cover, kickstart and gearchange pedals

1 Remove the gearbox outer cover by unscrewing the five Phillips screws, a single domed nut and a plain nut holding the outer cover to the inner cover. Pull away the cover complete with kickstart pedal assembly and gearchange mechanism. Take a note of the position of the screws as they vary in length.

3 Dismantling the engine/gearbox - removing the gearbox inner cover, selector quadrant and mainshaft

1 To remove the gearbox inner cover it is necessary to unscrew two cheese-head screws, one Allen screw and two bolts. This will enable the cover to be extracted with its associated selector quadrant and mainshaft.

2 The layshaft has a thrust washer located on the inside face of the inner cover by a small peg. A similar washer is fitted at the other end of the layshaft.

3 Unscrew the hexagon plug from the base of the gearbox housing. This contains a spring loaded plunger which operates in the cam plate notches.

4 It is now possible to remove the selector fork spindle and the large layshaft low gear followed by the sliding gears and their

selector forks. Special care is needed with the fork rollers which can easily be lost. The gearbox mainshaft can be withdrawn easily after the selector fork spindle has been removed.

5 Extract the layshaft with its fixed top gear and then the gearchange cam plate which has a spindle fitting into a boss on the wall of the housing.

6 To remove the gearbox top gear bearing it is necessary to detach the final drive sprocket as detailed in Chapter 1, Section 11 - 3.

7 The gear can be driven out of the bearing using a hammer and an alloy drift. It should be driven out from the outside into the inside of the gearbox. It is necessary to heat up the crankcase locally to the bearing prior to removal. This can be achieved by using cloths soaked in boiling water.

4 Dismantling the gearbox - removing and replacing the mainshaft and layshaft bearings

1 The mainshaft ball bearings are a press fit into their housings with circlips fitted to prevent sideways movement due to end thrust. If it is found necessary to replace them, proceed as follows:

2 To remove the right-hand bearing it is necessary after removing the circlip to heat the cover to about 100º C. Using a suitable sized drift knock out the old bearing and if you have a new one ready, drift it back in at once. Replace the circlip.

3 The high gear bearing is on the left of the machine. Remove the screws and oil seal holder.

4 Heat the casing in the bearing area to about 100º C and drift out the bearing from the inside of the casing. Drive in the replacement bearing immediately. Replace the circlip and then the oil seal and housing.

5 The layshaft right-hand needle roller bearing requires the cover heated to about 100º C to allow the bearing to be pressed out. Immediately fit in the replacement needle bearing from the inside of the cover. The bearing should protrude 0.073 - 0.078 inches above the cover face.

6 Access to the left-hand needle roller bearing is from the left, through the sprocket cover plate aperture. Heat the case to approximately 100º C and use a soft metal drift to drive the bearing into the gearbox. Fit the new bearing immediately until 0.073 - 0.078 inches protrudes above the surface inside the gearbox. The outer portion of the bore into which the bearing fits should be sealed.

7 The mainshaft high gear bush if necessary to renew, must be drifted out from the toothed end of the gear. The new bush must be pressed in with the oil groove in the bore of the bush at the toothed end. Take care to fit the correct type of bush.

8 The demonstration model and pictures of the stripdown in this manual are for the optional five speed gearbox. This was not available for fitment to the earlier Triumph and BSA machines.

9 The standard fitment on models up to 1973 however, is a four speed gearbox and the manual text is based on such a unit. A modification from a four to five speed gearbox is a task for the dealer as up to nearly forty new items are required in the kit.

10 Removal and refitment of the gearbox and associated components are similar but when refitting the quadrant the alignment procedure will change as per the accompanying illustrations.

5 Examination and renovation - general

1 Before the gearbox is reassembled, it will be necessary to inspect each of the components for signs of wear or damage. Each part should be washed in a petrol/paraffin mix to remove all traces of oil and metallic particles which may have accumulated as the result of general wear and tear within the gearbox.

2 Do not omit to check the castings for cracks or other signs of damage. Small cracks can often be repaired by welding, but this form of reclamation requires specialist attention. Where more extensive damage has occurred it will probably be cheaper to purchase a new component or to obtain a serviceable secondhand part from a breaker.

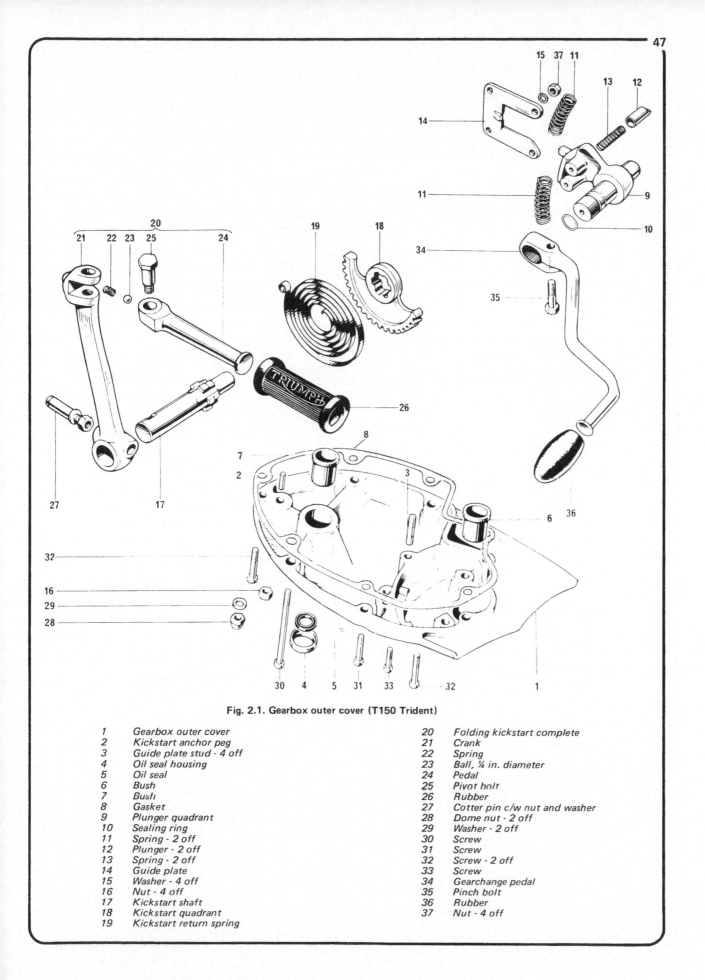

Fig. 2.1. Gearbox outer cover (T150 Trident)

1	Gearbox outer cover	20	Folding kickstart complete
2	Kickstart anchor peg	21	Crank
3	Guide plate stud - 4 off	22	Spring
4	Oil seal housing	23	Ball, ¼ in. diameter
5	Oil seal	24	Pedal
6	Bush	25	Pivot bolt
7	Bush	26	Rubber
8	Gasket	27	Cotter pin c/w nut and washer
9	Plunger quadrant	28	Dome nut - 2 off
10	Sealing ring	29	Washer - 2 off
11	Spring - 2 off	30	Screw
12	Plunger - 2 off	31	Screw
13	Spring - 2 off	32	Screw - 2 off
14	Guide plate	33	Screw
15	Washer - 4 off	34	Gearchange pedal
16	Nut - 4 off	35	Pinch bolt
17	Kickstart shaft	36	Rubber
18	Kickstart quadrant	37	Nut - 4 off
19	Kickstart return spring		

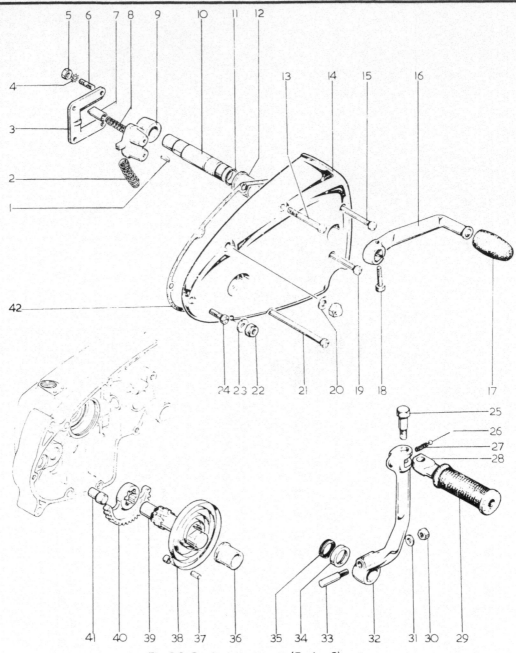

Fig. 2.2. Gearbox outer cover (Rocket 3)

1	Key	22	Nut
2	Spring - 2 off	23	Washer
3	Plate	24	Screw
4	Washer - 4 off	25	Bolt
5	Nut - 4 off	26	Steel ball
6	Stud - 4 off	27	Spring
7	Gearchange quadrant plunger - 2 off	28	Kickstart crank folding pedal
8	Gearchange quadrant spring - 2 off	29	Rubber
9	Gearchange plunger quadrant	30	Nut
10	Gearchange quadrant spindle	31	Washer
11	'O' ring	32	Kickstart crank
12	Bush	33	Kickstart crank cotter
13	Screw	34	Oil seal housing
14	Outer cover	35	Oil seal
15	Screw - 2 off	36	Bush
16	Gearchange pedal	37	Anchor pin
17	Rubber	38	Spring
18	Bolt	39	Spindle
19	Dome nut	40	Kickstart quadrant
20	Washer	41	Stop peg
21	Screw	42	Gasket

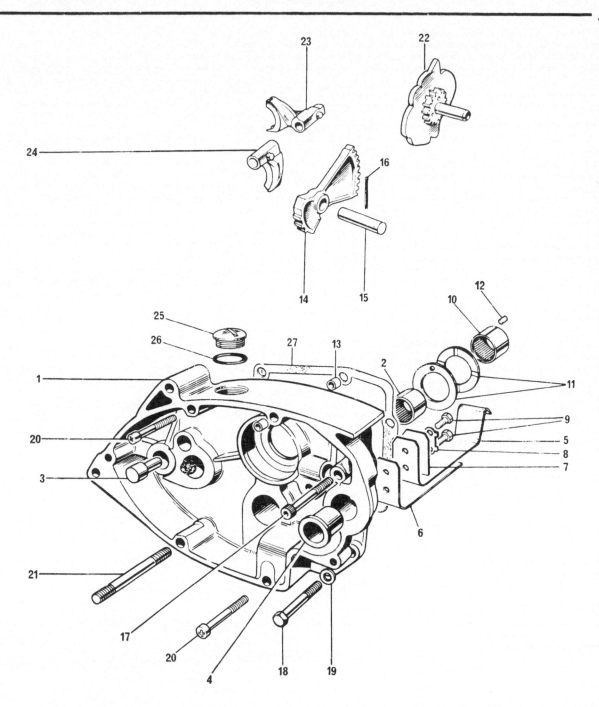

Fig. 2.3. Gearbox inner cover and selectors

1	Inner cover	15	Quadrant spindle
2	Layshaft bearing	16	Split pin
3	Kickstart stop	17	Socket screw
4	Bush	18	Bolt
5	Leaf spring	19	Washer - 2 off
6	Support plate	20	Screw - 2 off
7	Top plate	21	Stud - 2 off
8	Tab washer	22	Camplate
9	Screw - 2 off	23	Mainshaft selector fork
10	Layshaft bearing	24	Layshaft selector fork
11	Thrust washer - 2 off	25	Filler cap
12	Peg	26	'O' ring
13	Dowel - 2 off	27	Gasket
14	Quadrant		

3   If there is any doubt about the condition of a part examined, especially a bearing, it is wise to play safe and renew. A considerable amount of stripdown work will be required again if the part concerned fails at a much earlier date than anticipated.

6   Kickstarter mechanism - examination, renovation and replacement

1   Remove the kickstarter pedal from its shaft by loosening the cotter pin nut and tapping out the cotter pin itself using a soft metal drift. When the pedal is off its shaft it is possible to withdraw the kickstart quadrant and spring assembly from the inside of the cover.
2   If the engine/gearbox is still fitted in the machine the rear brake can be applied to hold the gearbox mainshaft to permit removal of the kickstarter ratchet pinion securing nut. The tab washer locking tab must be bent back at right angles first of all. Withdraw the pinion, ratchet, spring and sleeve for examination.
3   Thoroughly clean all parts in paraffin and check for damage, wear and cracks etc.
4   If the kickstarter quadrant shows signs of wear it should be renewed, otherwise it will tend to jam during the initial engagement with the ratchet pinion. The first tooth of the quadrant is relieved to minimise the risk of a jam when the initial contact is made.
5   If the kickstarter ratchet has to be renewed it is probable that the ratchet pinion with which it engages will also need renewal. It is not good engineering to run old and new parts together.
6   Examine carefully the ratchet teeth and renew both parts of the ratchet system if the teeth are rounded at their edges. A worn ratchet will eventually slip and make starting difficult. Renew the light return spring if it has taken a permanent set or has reduced return action.
7   If the kickstarter quadrant is to be renewed the spindle should be driven out using a hammer or press. The gear quadrant must be pressed onto the spindle so that the kickstarter crank location flat is positioned correctly relative to the quadrant or the operating angle will be incorrect. Fit a new oil seal over the shaft in the outer face of the end cover.
8   To reassemble proceed as follows:
     Fit the return spring to the kickstarter quadrant. Insert the spindle into the kickstarter bush and locate the return spring onto the anchor peg at the rear of the cover. Fit the oil seal over the spindle and assemble the kickstarter crank locking it into position with the cotter pin from the rear.
9   Refit the thin walled steel sleeve, spring pinion and ratchet to the gearbox mainshaft.
10 Fit a new tab washer in position and then screw on the retaining nut. It is essential that this is tightened to 40 - 45 ft lbs (5.5 - 6.2 kg m) and not overtightened or the thin walled inner steel sleeve will be damaged.
11 Turn over the locking tab on the tab washer.
12 Refit the outer cover and refill with the correct grade of lubricant.

7   Gearchange mechanism - examination and renovation

1   It should not be necessary to dismantle the gearchange mechanism unless the gear lever return springs have broken or wear of the operating mechanism is suspected on account of imprecise gear changes. To dismantle the assembly, detach the gear change lever by slackening the pinch bolt and pulling the lever off the splined shaft. Remove the four nuts and lockwashers securing the guide plate to the inside of the outer end cover and withdraw the guide plate complete with plunger quadrant and the curved gearchange lever return springs.
2   Examine the various components for wear, especially the gear change plungers and the plunger springs. Each spring should have a free length of 1¼ in; renew them if they have taken a set. The plungers must be a clearance fit in the quadrant if they are to function correctly.

3   If the plunger guide plate is worn or grooved on the taper guide surfaces it must be renewed.
4   The gearchange lever return springs seldom give trouble unless they become fatigued or if condensation within the gearbox causes corrosion. If there is any doubt about their condition renew them as a matter of course.
5   After a lengthy period of service the gearchange quadrant bush may wear oval. If this form of wear is evident, the bush must be renewed.
6   If the teeth of the camplate operating quadrant attached to the inner end cover are chipped, indented or worn, the quadrant must be renewed. It is retained by two split pins which, when removed, will permit the spindle to be withdrawn.
7   Do not omit to check the oil seal around the shaft of the gearchange lever in the outer end cover.

8   Gearbox components - examination and renovation

1   Examine each of the gear pinions carefully for chipped or broken teeth. Check the internal splines and bushes. Instances have occurred where the bushes have worked loose or where the splines have commenced to bind on their shafts. The two main causes of gearbox troubles are running with low oil, and condensation, which gives rise to corrosion. The latter is immediately evident when the gearbox is dismantled.
2   The mainshaft and the layshaft should both be examined for fatigue cracks, worn splines or damaged threads. If either of the shafts have shown a tendency to seize, discolouration of the areas involved should be evident. Under these circumstances check the shafts for straightness.
3   Harsh transmission is often caused by rough running ball races especially the mainshaft ball journal bearings. Section 5 of this Chapter describes the procedure for removing and replacing the gearbox bearings.
4   All gearbox bearings should be a tight fit in their housings. If a bearing has worked loose and has revolved in the housing, a bearing sealant such as Loctite can be used, provided the amount of wear is not too great.
5   Check that the selector forks have not worn on the faces which engage with the gear pinions and that the selector fork rod is a good fit in the gearbox housings. Heavy wear of the selector forks is most likely to occur if replacement of the mainshaft bearings is long overdue.
6   The gear selector camplate will wear rapidly in the roller

6.7 Locate the return spring onto the anchor peg

6.8 Refit the spring, pinion and ratchet

6.9 Tighten to 40 - 45 ft/lbs

6.10 Turn over the locking tab on the tab washer

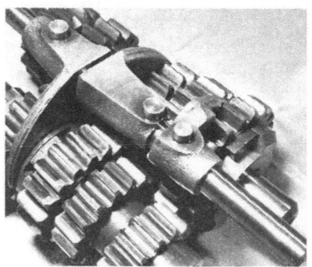

8.1 Examine each of the gear pinions carefully

8.2 Remove circlip to free pinion

8.6 The gear selector camplate will wear rapidly if the mainshaft bearings need replacement

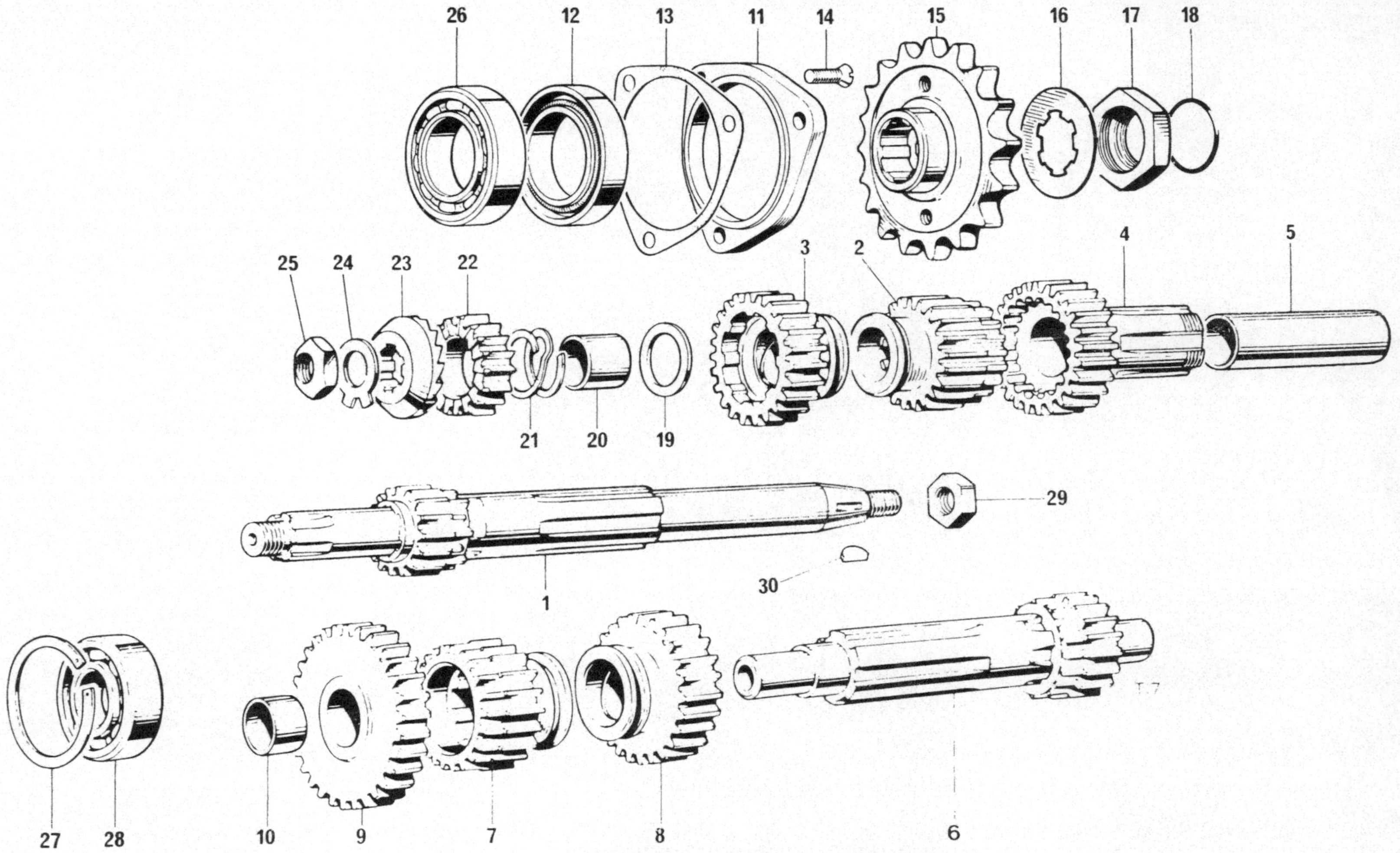

Fig. 2.4. Gearbox shafts and gears (4 speed gearbox)

1	Mainshaft c/w low gear (16T)	11	Oil seal housing
2	Mainshaft 2nd gear (20T)	12	Oil seal
3	Mainshaft 3rd gear (23T)	13	Gasket
4	Mainshaft high gear (26T)	14	Screw - 3 off
5	High gear bush	15	Gearbox sprocket 19T
6	Layshaft c/w high gear	16	Tab washer
7	Layshaft 3rd gear (22T)	17	Sprocket assembly locknut
8	Layshaft 2nd gear (26T)	18	Seal
9	Layshaft low gear (30T)	19	Kickstart pinion washer
10	Low gear bush	20	Kickstart pinion sleeve

21	Kickstart pinion spring		
22	Kickstart pinion		
23	Kickstart ratchet		
24	Mainshaft nut tab washer		
25	Mainshaft nut		
26	High gear bearing		
27	Circlip		
28	Mainshaft bearing		
29	Clutch hub to mainshaft nut		
30	Key		

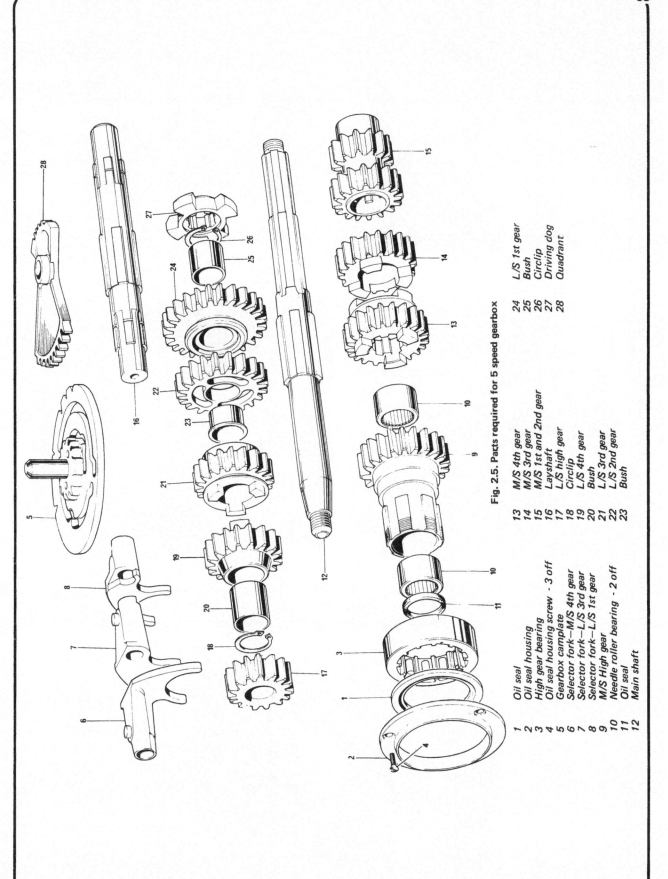

Fig. 2.5. Parts required for 5 speed gearbox

1 Oil seal
2 Oil seal housing
3 High gear bearing
4 Oil seal housing screw - 3 off
5 Gearbox camplate
6 Selector fork—M/S 4th gear
7 Selector fork—L/S 3rd gear
8 Selector fork—L/S 1st gear
9 M/S High gear
10 Needle roller bearing - 2 off
11 Oil seal
12 Main shaft

13 M/S 4th gear
14 M/S 3rd gear
15 M/S 1st and 2nd gear
16 Layshaft
17 L/S high gear
18 Circlip
19 L/S 4th gear
20 Bush
21 L/S 3rd gear
22 L/S 2nd gear
23 Bush

24 L/S 1st gear
25 Bush
26 Circlip
27 Driving dog
28 Quadrant

10.5 Assemble the camplate plunger as shown

10.9 Position the selector forks in their respective grooves

10.11 Locating the mainshaft and layshaft assemblies

11.3 Quadrant must be in this position

11.5 Temporarily replace the gearbox outer cover

tracks if the mainshaft bearings need replacement. Although the gear pinion behind the camplate is unlikely to wear excessively, it should be inspected if it has proved difficult to select gears.

7   The camplate plunger must work freely within its housing. Check the free length of the spring, which should be 2 21/32 in. if it has not compressed.

8   Play, accompanied by oil leakage, is liable to occur if the bush within the sleeve gear pinion is worn. The working clearance is normally from 0.003 in. to 0.005 in.

9   Gearbox reassembly - general

1   Before commencing reassembly, check that the various jointing surfaces are clean and undamaged and that no traces of old gasket cement remain. This check is particularly important because the gearbox itself is assembled with face to face joints and no gaskets. The gearbox will not remain oiltight if these simple precautions are ignored.

2   Check that all threads are in good condition and that the locating dowels, where fitted, are positioned correctly. Have available an oil can filled with clean engine oil so that the various components can be lubricated during reassembly.

10   Gearbox reassembly - replacing the camplate, camplate plunger, gear pinions and selectors

1   Fit the high gear bearing oil seal in its triangular housing, the closed side flush with the outer face. Press the high gear into the bearing.

2   Lubricate the ground tapered boss of the sprocket with oil and slide it onto the high gear. Screw on the securing nut finger tight.

3   Remesh the rear chain with the sprockets and replace the connecting link. It is essential to note that the closed end of the connecting link must point in the direction of chain travel. Apply the rear brake and tighten the sprocket securing nut as tight as possible.

4   Lubricate the camplate spindle and offer it into the spindle housing within the gearbox.

5   Assemble the camplate plunger and spring it into the extended hexagonal plunger retaining nut and screw it into position underneath the gearbox with its sealing washer in position.

6   Set the camplate with the plunger located in the notch between second and third gear.

7   Locate the bronze thrust washer over the inner needle roller bearing. Hold the thrust washer in position with grease; its grooved surface is towards the layshaft.

8   Fit the camplate rollers onto the selector forks and hold them in position with grease.

9   Assemble the complete gear cluster and position the selector forks in their respective grooves in the gears. Note that the one for the mainshaft cluster has the smaller radius.

10 The assembly is now ready to be offered into the gearbox housing.

11 As the mainshaft and layshaft are being located in their res-

pective bearings the gears should be slid into position and aligned so that the selector fork rollers locate in the roller tracks in the camplate and the bores for the selector forks are approximately aligned.

12 The selector fork spindle should be oiled and slid in through the selector forks with its shoulder end first so that it fully engages in the gearbox housing. No force should be used. The mainshaft selector fork will be in the innermost position.

13 Examine the camplate operating quadrant for free movement in the inner cover. Position the bronze layshaft thrust washer over the needle roller bearing in the inner cover, holding it in position with grease.

11 Gearbox reassembly - replacing the inner end cover and indexing the gears

1   Ensure all the moving parts in the gearbox are fully lubricated and apply a fresh coat of jointing compound to the gearbox jointing surface.

2   Fit the two locating dowels in position and offer the inner cover assembly to the gearbox.

3   With the cover approximately a quarter of an inch from the gearbox jointing, position the camplate quadrant in the middle point of its travel and then push the cover fully home. The middle tooth then aligns with the mainshaft centre line. The five speed gearbox on the demonstration model has chalk marks to show the alignment for FIRST gear.

4   Refit the socket screw, the cross headed screws and a bolt retaining the inner end cover to the gearbox shell.

5   It is expedient to temporarily replace the outer cover and check that the gearchange sequence is correct by operating the gearchange lever and turning the rear wheel in unison. It may be necessary to readjust the position of the camplate quadrant in relation to that of the preset camplate. Offer up the inner cover again ensuring that the middle tooth is on the mainshaft centre line.

12 Gearbox reassembly - replacing the outer end cover and completing assembly

1   Coat the jointing surfaces of the outer and inner end covers with a layer of gasket cement. Fit the two location dowels and turn the kickstart pedal until it is halfway down its operational stroke and offer the outer cover to the gearbox. Ensure that the kickstart pedal returns to its normal fully returned position.

2   Replace the top and bottom hexagonal nuts and the recess screws in the periphery of the outer cover.

3   Check that all the gears are selected in the correct sequence. It will be necessary to turn the rear wheel when making this check to ensure the gear pinions engage to their full depth.

4   Ensure the drain plug and level plug are refitted.

5   Refill the gearbox with the correct grade and quantity of oil as given in the Specifications Section at the beginning of the manual.

Fault diagnosis: following page

13 Fault diagnosis: gearbox

Symptom	Cause	Remedy
Difficulty in engaging gears	Gears not indexed correctly	Check timing sequence of inner end cover (will occur only after rebuild).
	Worn or bent gear selector forks	Examine and renew if necessary.
	Worn camplate	Examine and renew if necessary.
	Low oil content	Check gearbox oil level and replenish.
Machine jumps out of gear	Mechanism not selecting positively	Check for sticking camplate plunger or gear change plungers.
	Sliding gear pinions binding on shafts	Strip gearbox and ease any high spots.
	Worn or badly rounded internal teeth in pinions.	Replace all defective pinions.
Kickstarter does not return when engine is started or turned over	Broken kickstarter return spring	Remove outer end cover and replace spring.
	Kickstarter ratchet jamming	Remove end cover and renew all damaged parts.
Kickstarter slips on full engine load	Worn kickstarter ratchet	Remove end cover and renew all damaged parts
Gear change lever fails to return to normal position	Broken or compressed return springs	Remove end cover and renew return springs.

Chapter 3 Clutch

Contents

Specifications

Clutch

Type	Borg and Beck single dry plate
Overall thickness of friction plate inches	0.262
Overall thickness of friction plate mm	6.654
Diaphragm spring (maximum release load) lbs.	1000
Diaphragm spring (maximum release load) kgm	453.6
Minimum travel to disengage inches	0.035
Minimum travel to disengage mm	0.889
Minimum wear of friction plate inches	0.06
Minimum wear of friction plate mm	1.524
Bearing - outer thrust plate inches	½ x 1 1/8 x ¼
Bearing - outer thrust plate mm	12.7 x 28.575 x 6.35
Needle race (2 off) inches	1 3/8 x 1 5/8 x ½
Needle race (2 off) mm	34.93 x 41.28 x 12.7
Thrust race inches	1 3/8 x 2 1/16 x 5/64
Thrust race mm	34.93 x 52.39 x 1.984

1 General description

1 The clutch fitted to the Triumph Trident and the BSA Rocket 3 is the Borg and Beck dry single plate diaphragm spring unit.
2 Power output from the engine is transmitted from the engine sprocket by a triple row primary chain to the clutch chain wheel. Thence through the rubber shock absorber to the clutch. From the clutch the power is transmitted through the gearbox to the gearbox high gear and the gearbox final drive sprocket and thence by the rear chain to the wheel.
3 The task of the clutch is to provide the means of disconnecting this power train.
4 The main components of the clutch are a diaphragm spring, a cast iron pressure plate, a driven plate. These are enclosed in a pressed steel cover bolted to which is a cast iron drive ring.
5 The clamping load on the pressure plate is caused by the diaphragm spring.
6 The clutch release is due to a pull rod acting on a ball bearing in the centre of the pressure plate. This separates the two friction surfaces to allow the driven plate to operate freely between them.

2 Clutch mechanism - dismantling, renovation and reassembly

1 When the clutch is dismantled for replacement of a component due to unsatisfactory clutch operation or for a complete over-

haul is an opportune time to examine all clutch components for wear or damage.
2 Before stripping the clutch, mark the cover, drive ring and pressure plate to ensure reassembly in the identical position.
3 Bend back the tabs on the six tab washers holding the 12 end cover bolts in position and release the bolts half a turn at a time around the cover until the spring pressure is released when they can be removed. They must not be removed until the spring load is fully released. Failure to observe this precaution may cause permanent distortion of the cover.
4 Note there are three dowels in the drive ring which should not drop out and be lost.
5 Lift out the cast iron pressure plate complete with its bearings.
6 Before removing the driven plate ensure that there is clean brown paper on the bench and that the hands are clean. No lubricant is allowed on the driven plate facings.
7 Examine the driven plate rivets for condition. These should be below the surface of the facings. Do not attempt to lower the rivets with a punch as this will distort the driven plate and cause clutch drag. Renew contaminated facings. Ensure all rivets are tight.
8 Ensure that the splines are a smooth sliding fit on the clutch hub splines. If in any doubt replace the clutch hub or driven plate. Don't forget to check the teeth of the starter pinion on the clutch body.

2.3 Bend back the tabs on the six tab washers

2.10 The slots in the drive ring should not be damaged

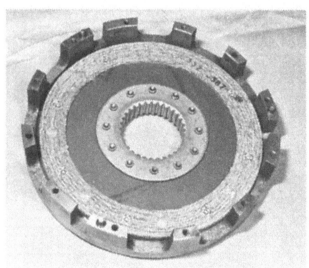

2.14 Place the driven plate in position

2.16 Slide the pressure plate lugs into position

2.18A Replace the cover in position again

2.18B Engage the three ribs of the cover around the spring

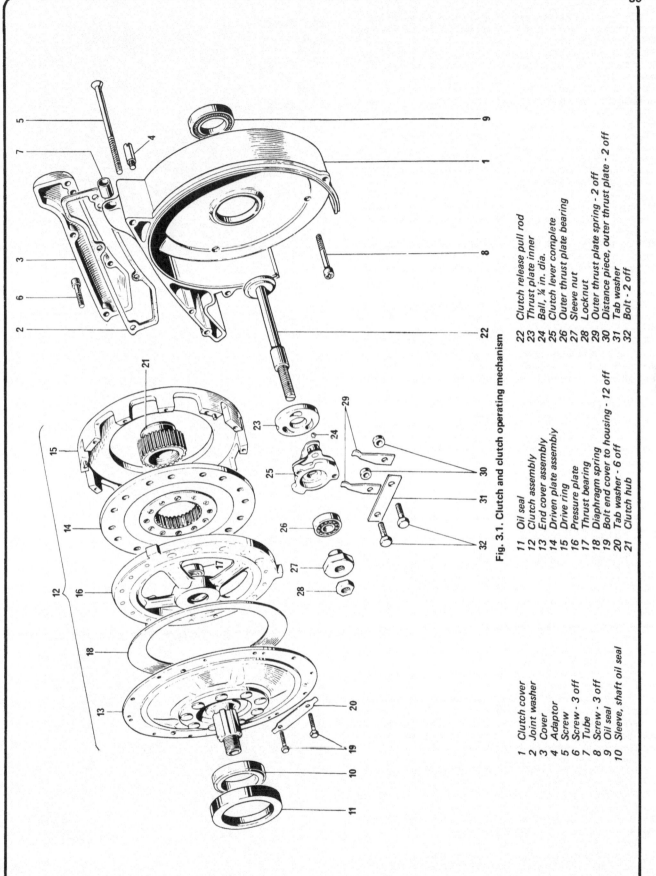

Fig. 3.1. Clutch and clutch operating mechanism

1 Clutch cover
2 Joint washer
3 Cover
4 Adaptor
5 Screw
6 Screw - 3 off
7 Tube
8 Screw - 3 off
9 Oil seal
10 Sleeve, shaft oil seal
11 Oil seal
12 Clutch assembly
13 End cover assembly
14 Driven plate assembly
15 Drive ring
16 Pressure plate
17 Thrust bearing
18 Diaphragm spring
19 Bolt end cover to housing - 12 off
20 Tab washer - 6 off
21 Clutch hub
22 Clutch release pull rod
23 Thrust plate inner
24 Ball, ¼ in. dia.
25 Clutch lever complete
26 Outer thrust plate bearing
27 Sleeve nut
28 Locknut
29 Outer thrust plate spring - 2 off
30 Distance piece, outer thrust plate - 2 off
31 Tab washer
32 Bolt - 2 off

9   The pressure plate central bearing should be examined for wear. Ensure no lubricant from the oil seal or bearing leaks onto the facings in the process.

10 The slots in the drive ring and the lugs on the pressure plate should not be damaged or burred. A loose fit can mean clutch out of balance.

11 Examine the diaphragm spring to detect discolouration due to overheating. If this is found then the clutch has been slipping. This could be due to weakness of the diaphragm which needs renewing.

12 Great care is needed for clutch reassembly so proceed as follows:

13 Place the clutch housing on the bench, now fitted with clean brown paper, with the housing friction surface uppermost.

14 Place the driven plate in position with the splines extending downwards.

15 Apply a light smear of high melting point grease to the sides of the three pressure plate lugs - don't drop any on the friction surface.

16 Slide the pressure plate lugs into position ensuring that the locating marks previously made are now aligned.

17 Apply a light smear of high melting point grease to the pressure plate machined ridge and fit the diaphragm spring with its outer edge upwards.

18 Lightly grease the inside edge of the cover and replace it in position once again referring to the alignment marks. Engage the three ribs of the cover around the spring.

19 Ensure when fitting new locking plates that the dowels in the clutch cover are between and not beneath the tab plates.

20 Insert the twelve end cover bolts at finger tightness and ensure the clutch plates are fully centralised before tightening the bolts.

21 Tighten the twelve bolts one half turn at a time around the clutch until the cover and housing meet.

22 Insert the clutch pull rod before refitting the clutch to the engine and smear the driven plate splines lightly with high melting point grease and slide the clutch onto the mainshaft and clutch hub. Secure with the hub nut.

2.22 Slide the clutch on to the mainshaft

3.1 The semi-spherical ramps

3 Clutch operating mechanism - dismantling, renovation, reassembly and resetting

1   The clutch is operated on the Trident and Rocket 3 machines by the rider pulling the left-hand handlebar lever towards himself. The movement is transmitted via the clutch cable to the clutch lever in the primary case. This clutch lever revolves on a bearing. Between the lever and the outer thrust plate there are three ¼ inch diameter steel ball bearings located in semi-spherical ramps. With the lever operated the balls roll up these ramps, forcing the two plates apart and operating the pull rod. The rod pulls the pressure plate outwards. This compresses the diaphragm spring and frees the driven plate.

2   The clutch operating mechanism should not need attention unless suspected to be faulty.

3   To disassemble bend back the tab washer from the two bolts retaining the spring clips and remove the tab washer, springs and spacers. Next remove the clutch lever which will release the three steel balls and lift off the thrust plate. Examine all parts carefully and replace if necessary. Clean all parts thoroughly.

4   Reinstall as follows:
   Refit the thrust plate into its register in the primary case taking care to locate on the dowel.

5   Smear the ramps with grease and fit the balls into the deepest parts of the ramps.

6   Fit the clutch lever with bearing over the balls ensuring that the cable trunnion is at 3 o'clock (45º angle).

7   Refit the spring clips and new tab washer.

8   Screw the large adjuster nut onto the pull rod turning the lever until the steel balls are at the lowest position in their ramps.

9   Slacken off the adjuster at the handlebars completely.

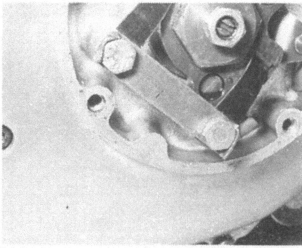
3.3 Bend back the tab washer from the two bolts

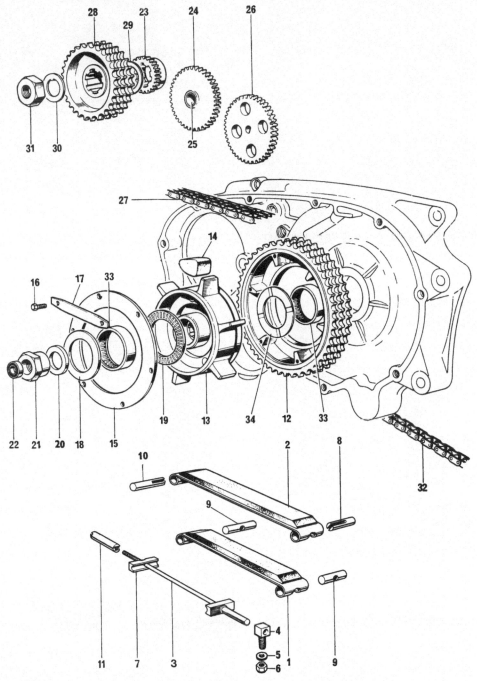

Fig. 3.2. Primary transmission and oil pump drive

1	Tensioner blade	18	Outer thrust washer
2	Tensioner blade	19	Thrust race
3	Tie rod	20	Spacer
4	Eye bolt	21	Nut
5	Washer	22	Oil seal
6	Nut	23	Driven gear
7	Saddle washer - 2 off	24	Intermediate gear c/w bush
8	Trunnion	25	Bush
9	Trunnion - 2 off	26	Oil pump driven gear
10	Pivot bolt	27	Primary chain
11	Adjuster nut	28	Engine sprocket (28T)
12	Clutch sprocket (50T)	29	Engine sprocket shim
13	Spider	30	Tab washer
14	Shock absorber rubber - 12 off	31	Engine sprocket nut
15	Retaining plate	32	Rear chain
16	Bolt - 6 off	33	Needle race - 2 off
17	Tab washer - 3 off	34	Inner thrust washer

3.6 Ensure the cable trunnion is at 3 o'clock

4.1A The circular plate inside the clutch chainwheel

4.1B Prising out the 12 individual rubbers

4.2 When refitting do not use grease or oil

4.3 Turn up the tab washers against the bolt lead flats

5.1 The chain itself is semi-adjustable

10 Insert a feeler gauge with a 0.005 inch between the bearing and the large adjuster nut.

11 Replace the small locknut and with a screwdriver blade inserted into the slot in the push rod tighten the locknut.

12 Adjust the cable to give 1/16 inch free play at the handlebar lever.

13 Operate the clutch to ensure correct action.

4 Examining the clutch shock absorber assembly

1 Inspection of the shock absorber rubbers necessitates removing six countersunk screws from the circular plate inside the clutch chain wheel and prising out the 12 individual rubbers.

2 These should be 100% sound with no indication of disintegration. When refitting do not use grease or oil as this will harm the rubber. If lubrication is necessary to refit then liquid soap is recommended. Apply a small amount of 'Loctite' grade A.V. to the six countersunk screws securing the circular screws on replacement. It is possible to caulk the heads in position with a centre punch.

3 There have been modifications to the size and shape of the twelve rubbers and Triumph and BSA types also differ. On later BSA models due to the six countersunk screws working loose, these have been replaced by six hexagon head bolts and three tab washers. A new plate without countersunk holes is also needed. Turn up the tab washers against the bolt head flats.

4 Modification kit numbers (BSA) are:
Plate   57.4004
Bolts   57-3940
Tab washers   57-3941

5 Adjusting the primary chain

1 The primary chain fitted to the Trident and Rocket machines is the triple row type. The chain itself is non-adjustable as the engine and gearbox mainshafts it connects are themselves fixed.

2 After periods of use the primary chain will need adjusting and provision for this is made by the fitting of a rubber faced tension slipper blade beneath the chain. Before adjustment place a drip tray beneath the front of the case.

3 Adjust the chain with the engine stopped. First remove the slotted inspection plug and feel any free movement in the chain with the finger. Release the hexagon plug from the front of the outer cover.

4 The chain should be adjusted to 3/8 inch (0.9 cm) free movement by inserting a screwdriver into the head of the slotted adjuster. Turn clockwise to increase the tension and vice versa. This adjustment figure is for post 1971 Triumph models. If your machine is earlier check your operator's manual, as some earlier machines had free movements of ½ inch (1.2 cm). This was applicable for earlier Triumph Tridents.

5 When the correct tension is achieved replace the adjuster plug and the slotted inspection plug. Refilling with oil is unnecessary as restarting the engine will top up the chaincase to the correct level.

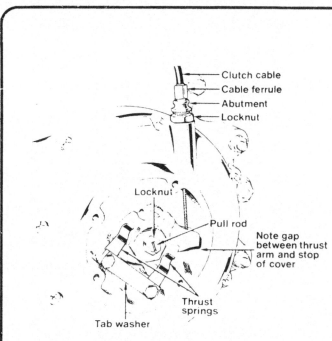

Clutch cable
Cable ferrule
Abutment
Locknut

Locknut

Pull rod

Note gap between thrust arm and stop of cover

Thrust springs

Tab washer

Fig. 3.3. Clutch adjustment

6   Fault diagnosis: clutch

Sympton	Cause	Remedy
Engine speed increases but machine does not respond	Clutch slip	Check clutch adjustment. If correct, suspect worn linings and/or weak springs.
Difficulty in engaging gears. Gear changes jerky and machine creeps forward, even when clutch is withdrawn fully	Clutch drag Clutch assembly loose on mainshaft	Check adjustment for too much play. Check tightness of retaining nut. If loose fit new tab washer and retighten.
Operating action stiff	Damaged, trapped or frayed control cable cable bends too acute	Check cable and replace if necessary. Re-route cable to avoid sharp bends.
Harsh transmission	Worn chain and/or sprockets	Replace.
Transmission surges at low speeds	Worn or damaged shock absorber rubbers	Dismantle clutch shock absorber and renew rubbers.

Chapter 4 Fuel system and lubrication

Contents

Specifications

Carburettor		T150	A75
Amal type	... ...	626	626
Number	... ...	626/48(L/H) 49(C) 47(R/H)	626/14, 15, 16
Bore	... ...	27 mm	27 mm
Main jet	... ...	150	150
Pilot jet	... ...	622/107	622/107
Needle jet	... ...	0.106	0.106
Needle position	... ...	2nd groove	2
Throttle valve	... ...	2½ (3½ '71 on)	3
Needle type	...	Std	Std
Air cleaner type	... ...	Filter cloth and wire gauze	Zig-zag felt
Fuel tank capacities	... ...	4.5 Imp gall	4 Imp galls
		20.4 litres	18.0 litres
Octane rating	... ...	97 Premium	As T150
Min	... ...	B 54050 - 4 star	As T150

Oil pressure readings		All models
Normal	... ...	75/85 p.s.i
		(5.273 - 5.624 kg/sq cm)
Idling ...	... ...	20/25 p.s.i.
		(1.406 - 1.758 kg/sq cm)
Oil pressure switch working range ...	... ...	7 - 11 lbs (3.178 - 4.994 kgm)
Oil pressure release valve		
Piston diameter	... ...	0.5625/0.5620 ins
		(14.287/14.2748 mm)
Working clearance	... ...	0.001/0.002 ins
		(0.0254/0.0508 mm)
Operating pressure	... ...	90 p. s.i.
		(6.328 kg/sq cm)
Spring length (free) ...	... ...	1.375 ins
		(34.925 mm)
Load at 1.1875	... ...	8 lbs
		(3.632 kg)
Rate	... ...	42.3 lb
		(19.2042 kg)

Oil pump

	T150	A75
Bore diameters	.34381/.3433 ins (8.7325/8.7198 mm)	
Bore diameter - scavenge gear	.3438/.3448 ins (8.7325/8.7579 mm)	
Bore diameter feed gear	As scavenge	
Spindle diameter	.3433/.3428 ins (8.7198 - 8.70712 mm)	
Cover plate bore Diameter : spindle	.3833/.3438 ins (8.7198 - 8.7325 mm)	
Cover plate bore Diameter drive scavenger gear	.4375/.4370 ins (11.1125/11.0998 mm)	
Pump drive ratio	1.9 : 1 (engine to pump)	As T150

1 General description

The fuel system comprises a petrol tank from which petrol is fed by gravity to the float chamber(s) of the carburettor(s). Two petrol taps, with built-in gauze filter, are located one each side beneath the rear end of the petrol tank. For normal running the right-hand tap alone should be opened except under high speed and racing conditions. The left-hand tap is used to provide a re-serve supply, when the main contents of the petrol tank are ex-hausted.

For cold starting the carburettor(s) incorporate an air slide which acts as a choke controlled from a lever on the handlebars. As soon as the engine has started, the choke can be opened gradually until the engine will accept full air under normal run-ning conditions.

Lubrication is effected by the 'dry sump' principle in which oil from the separate oil tank is delivered by gravity to the mech-anical oil pump located within the crankcase. Oil is distributed under pressure from the oil pump through drillings in the crank-shaft to the big ends where the oil escapes and is fed by splash to the cylinder walls, ball journal main bearings and the other inter-nal engine parts. Pressure is controlled by a pressure release valve, also within the crankcase. After lubricating the various engine components, the oil falls back into the crankcase, where it is re-turned to the oil tank by means of the scavenge pump. A bleed-off from the return feed to the oil tank is arranged to lubricate the rocker arms and valve gear, after which it falls by gravity via the pushrod tubes and the tappet block, to the crankcase. It will be noted that the oil pump is designed so that the scavenge plunger has a greater capacity than the feed plunger, this is nec-essary to ensure that the crankcase is not flooded with oil, and that any oil drain-back whilst the machine is standing is cleared quickly, immediately the engine starts.

2 Petrol tank - removal and replacement

1 The petrol tank is secured to the frame by two studs under-neath the nose, one on each side. These studs project through two short brackets which are attached to the frame and are cushioned by rubber shock mounts. The tank is retained at the front by two self locking nuts and washers which thread onto the studs. The rear mounting takes the form of a lug welded to the rear of the tank which matches with a threaded hole in the top portion of the frame, close to the nose of the dual seat. Anchorage is pro-vided by a bolt which passes through shaped rubbers to provide a flexible mounting.

2 Some models have a central fitting and it is necessary to re-move the four crosshead screws which retain the padded styling strip and then remove the strip. It is then necessary to remove the central fixing nut and washer to free the petrol tank.

3 Before lifting the tank from the frame, disconnect the fuel pipes at the unions. Ensure the various shaped rubber mountings are not lost as they will be displaced when the tank is lifted.

4 When refitting the tank take special care to ensure that none

of the carburettor control cables is trapped or bent to a sharp radius. Apart from making control operation much harder there is a risk that the throttle could stick since there is a minimal clearance between the underside of the fuel tank and the top frame tube.

3 Petrol taps - removal and replacement

1 The petrol taps are threaded into inserts in the rear of the petrol tank, at the underside. Neither tap contains provision for turning on a reserve quantity of fuel. It is customary to use the right-hand tap only so that the left-hand tap will supply the re-serve quantity of fuel, unless the machine is used for high speed work or racing. In these latter cases, it is essential to use both taps in order to obviate the risk of fuel starvation.

2 Before either tap can be unscrewed and removed, the petrol tank must be drained. When the taps are removed each gauze filter, which is an integral part of the tap body, will be exposed.

3 When the taps are replaced, each should have a new sealing washer to prevent leakage from the threaded insert in the bottom of the tank. Do not overtighten; it should be sufficient just to commence compressing the fibre sealing washer.

4 Petrol feed pipes - examination

1 Plastic feed pipes of the transparent variety are used with a union connection to each petrol tap and a push-on fit at the car-burettor float chamber.

2 After lengthy service, the pipes will discolour and harden gradually due to the action of the petrol. There is no necessity to renew the pipes at this stage unless cracks become apparent or the pipe becomes rigid and 'brittle'.

5 Carburettors - removal

1 Both the Rocket 3 A75 and the Trident T150 models utilise Amal concentric carburettors. The conventional throttle spring is not fitted, the throttle return being by a scissor spring on the external throttle linkage. Air slides are fitted, controlled via a junction box from the air control handlebar lever. On later models this control is local to the carburettors.

2 The three carburettors are mounted on a manifold which houses the throttle linkage mechanism and the manifold is in turn connected to the three inlet ports by three rubber hoses which are retained by jubilee clips.

3 After removing the petrol tank release the left-hand side panel by unscrewing the plastic knob at the top of the panel and care-fully sliding the panel forward, off its two spigotted rubber mounting bushes at the rear.

4 Disconnect the throttle cable from the carburettor linkage

3.1 The petrol taps thread into the rear of the petrol tank

5.5 Withdraw carburettors as a unit, from left

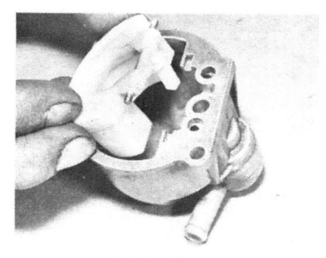

6.1 Lift out the horseshoe shaped float

6.8 Fit new gasket at float chamber joint if original is damaged

and also remove the air control cable from the handlebar lever.

5   Slacken off the worm drive clips at the carburettor end of the rubber inlet stubs and remove the rubber buffer from the air filter supporting bracket. The carburettor, manifold and air filter can now be withdrawn as a unit from the left-hand side of the machine

6   Each carburettor is removed from the manifold by unscrewing the two crosshead screws from each carburettor top and the two nuts from the holding studs.

7   The carburettor can then be drawn off the studs and downwards leaving the air and throttle slides in position on the manifold assembly.

8   To remove the carburettor top from the manifold, first disconnect the throttle slide from the throttle rod and the air slide from its cable. In the case of the throttle slide remove the needle retaining spring clip and compress the throttle rod spring so that the top retaining plate can be withdrawn. Push the bottom nipple of the throttle rod downwards clear of the throttle slide. Removal of the air slide necessitates only compression of the spring whilst the cable nipple is pushed clear of the slide. Unscrew the air cable abutment to free completely the carburettor top.

6   Carburettors - dismantling, examination and reassembly

1   To remove the float chamber, unscrew the two crosshead screws on the underside of the mixing chamber. The float chamber can then be lifted away complete with float assembly and sealing gasket. Remove the gasket and lift out the horseshoe shaped float, float needle and spindle on which the float pivots.

2   When the float chamber has been removed, access is available to the main jet, jet holder and needle jet. The main jet threads into the jet holder and should be removed first, from the underside of the mixing chamber. Next unscrew the jet holder which contains the needle jet. The needle jet cannot be removed until the jet holder has been unscrewed and removed from the mixing chamber because it threads into the jet holder from the top. There is no necessity to remove the throttle stop or air adjusting screws.

3   Check the float needle for wear which will be evident in the form of a ridge worn close to the point. Renew the needle if there is any doubt about its condition, otherwise persistent carburettor flooding may occur.

4   The float itself is unlikely to give trouble unless it is punc-

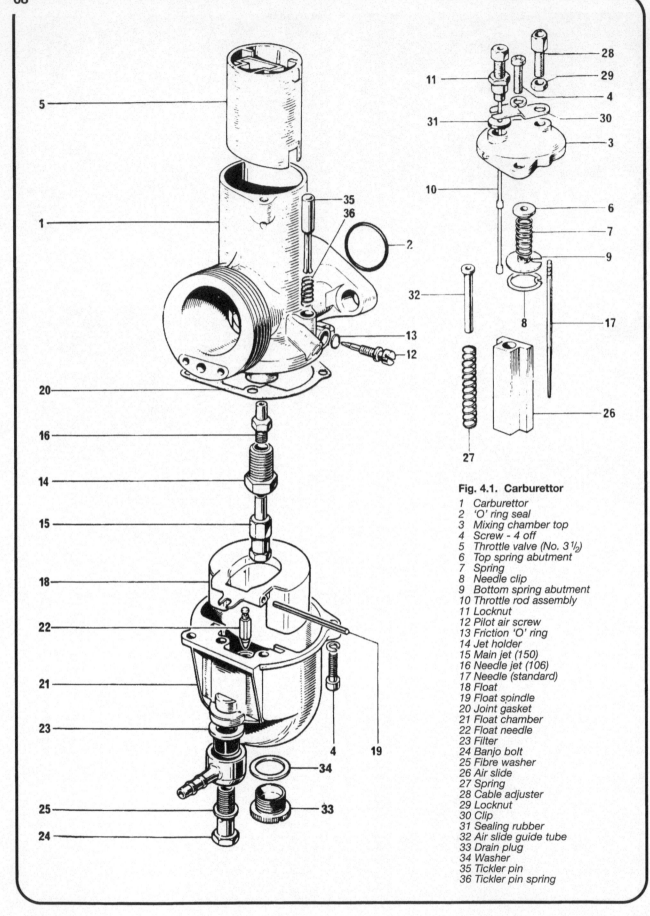

Fig. 4.1. Carburettor

1 Carburettor
2 'O' ring seal
3 Mixing chamber top
4 Screw - 4 off
5 Throttle valve (No. 3 1/2)
6 Top spring abutment
7 Spring
8 Needle clip
9 Bottom spring abutment
10 Throttle rod assembly
11 Locknut
12 Pilot air screw
13 Friction 'O' ring
14 Jet holder
15 Main jet (150)
16 Needle jet (106)
17 Needle (standard)
18 Float
19 Float spindle
20 Joint gasket
21 Float chamber
22 Float needle
23 Filter
24 Banjo bolt
25 Fibre washer
26 Air slide
27 Spring
28 Cable adjuster
29 Locknut
30 Clip
31 Sealing rubber
32 Air slide guide tube
33 Drain plug
34 Washer
35 Tickler pin
36 Tickler pin spring

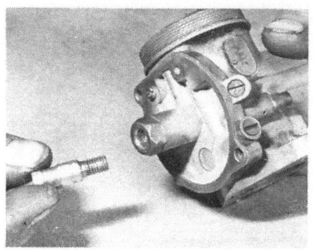

6.9 Check that the main jet and needle jet are clean

tured and admits petrol. This type of failure will be self-evident
and will necessitate renewal of the float.

5   The pivot needle must be straight - check by rolling the needle
on a sheet of plate glass.

6   It is important that the gasket between the float chamber and
the mixing chamber is in good condition if a petrol tight joint is
to be made. If it proves necessary to make a replacement gasket,
it must follow the exact shape of the original. A portion of the
gasket helps retain the float pivot in its correct location; if the pin
rides free it may become displaced and allow the float to rise,
causing continual flooding and difficulty in tracing the cause. Use
Amal replacements whenever possible.

7   Remove the union at the base of the float chamber and check
that the inner nylon filter is clean. All sealing washers must be in
good condition.

8   Make sure that the float chamber is clean before replacing the
float and float needle assembly. The float needle must engage
correctly with the lip formed on the float pivot; it has a groove
that must engage with the lip. Check that the sealing gasket is
placed OVER the float pivot spindle and the spindle is positioned
correctly in its seating.

9   Check that the main jet and needle jet are clean and un-
obstructed before replacing them in the mixing chamber body.
Never use wire or any pointed instrument to clear a blocked jet,
otherwise there is risk of enlarging the orifice and changing the
carburation. Compressed air provides the best means, using a tyre
pump if necessary.

10 Before refitting the float chamber, check that the jet holder
and main jet are tight. Do not invert the float chamber, otherwise
the inner components will be displaced as the retaining screws are
fitted. Each screw should have a spring washer to obviate the risk
of slackening.

11 When replacing the carburettor, check that the 'O' ring seal in
the flange mounting is in good condition. It provides an airtight
seal between the carburettor flange and the cylinder head flange
to ensure the mixture strength is constant. Do not overtighten the
carburettor retaining nuts for it is only too easy to bow the flange
and give rise to air leaks. A bowed flange can be corrected by removing the 'O' ring and rubbing down on a sheet of fine emery
cloth wrapped around a sheet of plate glass, using a circular
motion. A straight edge will show if the flange is level again, when
the 'O' ring can be replaced and the carburettor refitted.

12 Before the mixing chamber top is replaced, check the throttle
valve for wear. A worn valve is often responsible for a clicking
noise when the throttle is opened and closed. Check that the
needle is not bent and that it is held firmly by the clip.

13 Trident and Rocket 3 carburettors are fitted with a tickler
for use when starting with a dry carburettor.

7   Synchronising the triple carburettors

1   In order to synchronise the throttle slides the carburettor
assembly must be removed from the machine.

2   Site the carburettor and inlet manifold assembly less the air
filter on the workbench. View the throttle slides through the
engine side of the carburettor and adjust the throttle stop screw
until one slide is open approximately 0.010 inches. Compare the
other two slides and adjust the slide heights by screwing the individual adjusters (one on top of each carburettor) clockwise to
lower the slides, and anticlockwise to raise the slides. There is a
locknut on each adjuster and this should be tightened when the
adjustment is complete. The difference in slide heights is easily
visible due to the slide opening characteristics.

3   Turn each air screw fully in and unscrew them 2½ complete
turns to obtain the approximate fuel/air ratio at idle. Turn the
screw clockwise to richen the mixture and anticlockwise to weak-
en it.

4   Refit the carburettor assembly to the machine and adjust the
throttle stop screw to give an idle of approximately 500 rpm
when the engine is warm.

5   Care must be taken to see that all six carburettor connection
pipe clips are tightened.

6   On the Triumph Trident demonstration model illustrated in
this Chapter, the air control was fitted to the carburettor assembly and not to the lever on the right handlebar and eliminates the
need for a cable junction under the dual seat. The air control can
be checked on a cold engine by moving the lever forward to the
slack wire position. It should be opened progressively as the
engine warms. The control partially blocks the passage of air
through the main choke.

8   Air cleaner - removal and replacement

1   Remove both side panels. The right-hand side panel is retained by two 'Posidriv' screws and the left panel is secured by a
plastic knob. Withdraw both side panels to gain access to the air
filter unit.

2   Remove the two bolts securing the air filter to the carburettor
and then remove the breather pipe from the rear of the filter.

3   To remove the filter element remove the screw and claw holding the surrounding perforated band. Lift away the perforated
band and then the filter backplate.

4   The filter element can now be washed in petrol and dried
with air blown from a foot pump.

5   The front plate of the filter is secured to the carburettor
assembly by two long bolts at the carburettor mounting flanges.

6   Two types of filter element are employed on these machines.
One is a convoluted paper element and the other is formed from
cloth or felt backed with wire gauze. Neither of these elements
should come in contact with oil.

7   Reassembly is the reversal of the foregoing.

8   On no account run the machine with the air cleaner disconnected unless the carburettor has been rejetted to suit. The fitting of an air cleaner calls for the reduction in the size of the main
jet in order to compensate for the enriching effect of the air
cleaner element. If the air cleaner is left off a permanently weakened mixture will result. This could lead to early failure of the
valves and/or piston crowns.

9   Exhaust system - general

1   The exhaust manifold is a three into two device to transfer the
exhaust gases from the three cylinders into the two exhaust pipes.
A silencer is fitted to each exhaust pipe.

2   Slacken the two exhaust pipe to silencer clip bolts and the two
exhaust pipe to manifold pinch bolts. Tap downwards on the
exhaust pipes with a rubber mallet to remove them.

3 Slacken the bolts at the three manifold finned clips and re-move the exhaust manifold again using the rubber mallet to slide it from the exhaust stubs.

4 The right silencer can be lifted away from the clip bolted to the frame bracket.

5 Avoid unscrewing the adapter studs unnecessarily as damage to the threads in the cylinder head can be caused by engine vibration if they work loose.

10 Engine lubrication - checking the oil pressure

1 If the pump pressure in the lubrication system exceeds the permitted maximum a release valve in the left-hand crankcase base opens and passes the excess oil through to the bottom of the crankcase. With the engine inoperative oil pressure is nil but with a cold start a pressure of 90 p.s.i. is possible. At 3,000 rpm and with the engine hot this will fall to between 75 and 90 p.s.i. which is the operating pressure.

2 If there is a suspected fault with the oil pressure your dealer can insert an oil gauge connected to an adaptor into the hole created when a blanking plug on the front of the centre crank-case is removed. If this test is carried out by the owner and the correct pressure is not obtained the following checks must be carried out immediately.

3 The oil pressure relief valve must be clean and undamaged. Check the valve against the dimensions given in the Specifications Section of this Chapter.

4 Ensure that there is the full amount of lubricant in the oil tank and that the return circuit to the tank is operating.

5 Ensure that the filters for sump, crankcase and oil tank are functioning correctly and are clean.

6 Ensure that the crankcase filter is not fitted in reverse.

7 Ensure that the double gear type oil pump is operating effi-ciently and in turn is receiving the correct amount of oil.

8 Ensure that the drilled connections in the crankcase between oil pump and pipes are not blocked.

9 Ensure that the big-ends and centre plain main bearings have the correct working clearance and are not damaged or badly worn. If the latter is the case then oil can escape much more quickly and give a drop in indicated pressure.

10 There are rubber sealing rings at each of the main bearing to tappet guide blocks. Ensure that there is no oil leak at these points.

11 An overall drop in oil pressure can be caused by long sessions of slow running or excessive use of the air control. These can cause dilution in the oil tank.

11 Engine lubrication - removing and replacing the oil pump

1 The oil pump is driven by a gear train from the crankshaft. The moving parts are normally operating in the lubricant so that wear on them is minimal. It is essential however that after periods of long or very hard riding that both gears on the feed and scavenge sides of the pump are closely examined for wear. The pump is located on the drive side crankcase and protrudes through the inner primary chaincase. The pump body is made of cast iron.

2 To remove the pump proceed as follows: Remove the outer primary chaincase (see Chapter 1).

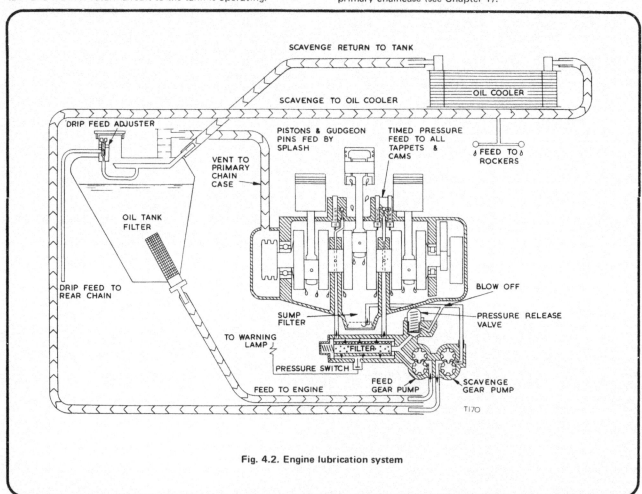

Fig. 4.2. Engine lubrication system

3 Remove the inner primary chaincase (see Chapter 1).
4 Remove the oil pump drive gear.
5 The oil pump is secured to the crankcase by four Pozidriv screws. Unscrew these items and the pump can be lifted clear.
6 Two slot head screws hold the three sections of the pump body and these must be removed for pump examination.
7 The two spindles can be gently tapped out from the crankcase side of the pump to release the gears. Wash all parts thoroughly in paraffin.
8 Dimensions for the oil pump components are given in the Specifications Section at the front of this Chapter. Compare the dimensions of the components against these and replace if they are out of tolerance.
9 The gear teeth should be examined under a lens for indentations and scuffing.
10 Reassembly is the reverse procedure from disassembly.
11 A new gasket will be required between the oil pump and gasket and ensure that the pump is correctly located over the dowel in the crankcase recess.
12 Ensure that the slot headed screws are fitted with serrated washers. Do not overtighten the Pozidriv mounting screws.
13 Fit a new 'O' ring into the chaincase facing recess surrounding the pump body.
14 Replace the transmission and chaincases as described in Chapter 1.
15 Do not dismantle the oil pump unless absolutely necessary.

12 Engine lubrication - removing and replacing the main oil feed filter

1 The main feed oil filter should be replaced at each oil change.
2 Remove the hexagon headed cap from below the forward end of the gearbox outer cover and lift out the spring. The element can be withdrawn using pliers. Dispose of the used filter.
3 When refitting ensure that the end of the element with the hole goes in first otherwise there will be an oil blockage.
4 Ensure that the rubber sealing ring is attached to the end of the element.
5 Replace the spring and cap ensuring the sealing ring and a new fibre washer are seated correctly.
6 Refill the oil tank.

13 Engine lubrication - removal and replacement of the crankcase sump filter

1 The oil should be drained when the engine is warm as it will run much more easily and quickly.
2 During the run-in period it is necessary to change the oil much more frequently and this should be carried out at the following intervals:
    250 miles   (400 km)
    500 miles   (800 km) and 1,000 miles (1,500 km)
3 The running-in process causes a controlled amount of wear on the engine and minute particles of metal are collected in the oil. and particularly in the sump filter. When the engine is hot the majority of these particles are contained in the lubricant and can be drained out with it.
4 Check with your main dealer if your machine is new as he will carry out some of this work for you under the warranty terms.
5 Drain the oil as instructed in paragraph 1 and have a receptacle such as a drain tin or even an old plastic bowl beneath the sump to catch the residue of oil. It is advisable to have your machine on a stand.
6 At the bottom of the crankcase remove the six nuts and locking washers securing the sump plate and remove first the plate, then the two gaskets and the wire gauze filter. Allow the sump to drain for about ten minutes.
7 The filter should be washed in paraffin to remove all foreign matter and the gaskets must be renewed.
8 With the sump plate removed, ensure that the sump suction pipe is clear.

8.2 Remove the breather pipe from the rear of the filter

9.5 Avoid unscrewing the adapter stubs unnecessarily

13.6 Wire gauze filter behind sump plate

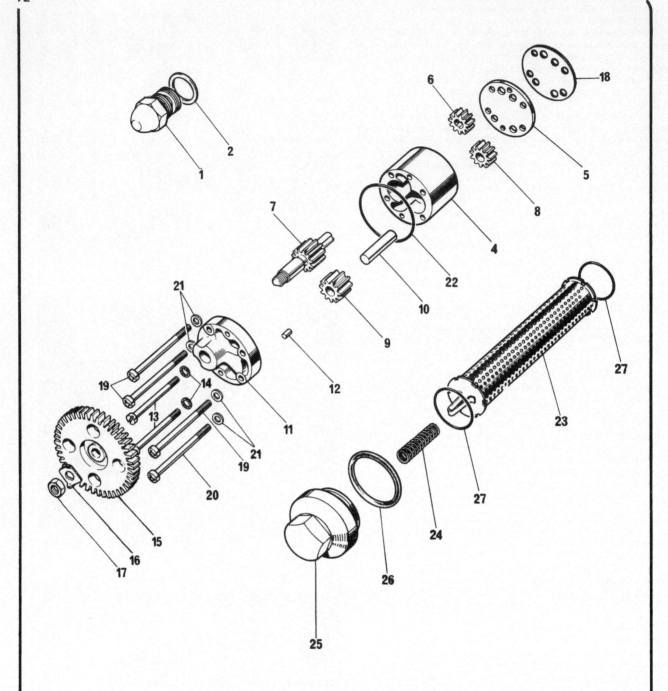

Fig. 4.3. Oil pump release valve and filter

1	Oil pressure release valve		15	Driven gear
2	Joint washer		16	Tab washer
3	Oil pump complete		17	Nut
4	Oil pump body		18	Gasket
5	Base plate		19	Screw - 3 off
6	Drive feed gear		20	Screw
7	Drive scavenge gear		21	Washer - 4 off
8	Driven feed gear		22	Seal
9	Driven scavenge gear		23	Oil filter element
10	Driven gear spindle		24	Spring
11	Cover		25	Cap
12	Cover to body dowel		26	'O' ring
13	Screw - 2 off		27	Filter seal
14	Washer - 2 off			

9 Refitting is the reverse procedure but note that there is a gasket on either side of the filter and that the pocketed end of the sump plate is towards the rear of the engine. Check for any leakage when the engine is running.

14 Engine lubrication - removing and replacing the oil pressure release valve

1 It is not anticipated that the oil pressure release valve should give trouble and require servicing other than periodic cleaning. The valve is located to the rear of the oil pump housing on the base of the engine.
2 Access to the valve is only possible after stripping the primary transmission. It is necessary to disconnect the oil feed and scavenge pipes at the crankcase. Tape up the ends of the pipes as they are disconnected.
3 With the valve removed the hexagonal domed cap can be unscrewed from the valve enabling the piston to be withdrawn.
4 All the parts of the valve should be cleaned in paraffin and thoroughly examined.
5 The piston should be checked for score marks and the spring for length and possible fracture. Dimensions are given in the Specifications Section at the beginning of this Chapter.
6 When reassembling fit new fibre washers at the valve cap and between the release valve body and the crankcase.

15 Engine lubrication - removing and replacing the oil cooler

1 It is necessary to remove the petrol tank in order to gain access to the oil cooler, clips and fittings. BSA models have embellished covers on the cooler to remove first.
2 Do not tilt the cooler until it is clear of the frame and can be drained into a receptacle or oil could be lost.
3 With the petrol tank removed unscrew the oil pipe clips. The left-hand pipe from the cooler connects to the rocker feed pipe.
4 The two support bracket top bolts should be slackened sufficiently to allow removal of the corner packings.
5 Take the weight of the cooler and remove the support bracket bolts plus their washers and nuts. Remember to collect the four spigotted rubber washers. Examine the rubber washers and check for cracks or perishing.
6 When the cooler has been drained do not attempt to flush out the interior. Clean the exterior with paraffin and brush.
7 Reassemble the support brackets after examination. The large oil pipe unions on the top of the cooler should face rearwards when refitting the cooler with brackets to the frame. Ensure the spigotted rubbers are correctly installed.

8 Replace the large diameter oil pipes. The left-hand pipe is the one connected to the scavenge pipe.
9 Tighten the pipe clips and refit the petrol tank.
10 Examine the reflector rubbers and plates fitted to the sides of the cooler for condition. The reflectors are attached to the rubber plates by screws and nuts but the rubber plates are attached to the cooler with rubber adhesive.

16 Engine lubrication - removing and replacing the oil tank filter

1 Remove the oil tank filter cap. The right-hand panel is removed by unscrewing the three Pozidriv screws securing it.
2 With a drain tray below the oil tank remove the drain plug and its fibre washer. Allow about ten minutes draining time.
3 Remove the union nut to disconnect the oil feed pipe and then unscrew the hexagon headed oil tank filter with its associated fibre washer.
4 Wash the filter thoroughly in paraffin and also the oil tank although flushing oil is preferable for washing out the tank.
5 Refitting is the reverse procedure to removal.
6 Refill with the correct grade of oil to the top line of the dipstick.

17 Engine lubrication - removing and replacing the rocker oil feed pipe

1 If it is necessary to remove the rocker oil feed pipe two domed nuts should be removed from the ends of the rocker spindle and the banjos withdrawn.
2 At the scavenger oil cooler pipe disconnect the rocker oil feed pipe and drain it into an oil tray. Take extreme care not to bend the pipe excessively as a fracture could occur.
3 Thoroughly clean the pipes with paraffin and blow through the banjos with an air line or foot pump to clear any blockage.
4 Refit the rocker oil feed pipes using new copper washers.

18 Engine lubrication - removing and replacing the anti-drain valve

1 This valve is fitted in the crankcase centre section adjacent to the oil pump housing and is used to prevent oil draining through from the feed side of the pump as could happen overnight and when the pump is well worn. This could also happen if the ball of the valve sticks and the crankcase would then fill with oil.
2 To service the valve remove the plug from the crankcase, making sure to receive the coil springs and ball in the hands. Clean in paraffin, examine and if intact, refit.

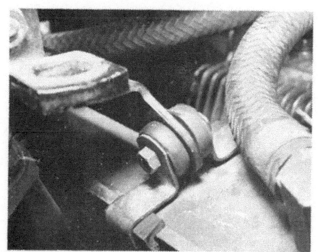

15.5 Oil cooler is suspended on rubber mountings

15.10 Make sure hose connections are tight

19 Fault diagnosis: fuel system and lubrication

Symptom	Cause	Remedy
Excessive fuel consumption	Air filter choked, damp or oily Fuel leaking from carburettor Float needle sticking Worn carburettors	Check and if necessary renew. Check all unions and gaskets. Float needle seat needs cleaning. Renew.
Idling speed too high	Throttle stop screw in too far Carburettor top loose	Re-adjust screw. Tighten top.
Engine does not respond to throttle	Mixture too rich	Check for displaced or punctured float.
Engine dies after running for a short while	Blocked air vent in filler cap Dirt or water in carburettors	Clean. Remove and clean float chambers
General lack of performance	Weak mixture; float needle stuck in seat Leak between carburettor and cylinder head Fuel starvation	Remove float chamber and check. Bowed flange; rub down until flat and replace O ring seal. Turn on both petrol taps for fast road work.

Chapter 5 Ignition system

Contents

Specifications

Ignition coils

Make	Lucas	Siba
Type	17-M112*	3200
Voltage	12v	

(Replace with 6 volt 17M6 for transistorised ignition)

Contact breakers

Make	Lucas	Advance range - 12° (24° crankshaft)
Type	7CA	
Gap ins.	0.014 - 0.016	Fully advanced at 2,000 rpm
Gap mm	0.35 - 0.40	

Spark plugs

Make and type	Champion N3	KLG FE 100	Lodge 2HLN	NGK B8ES
Size	14 mm	14 mm	14 mm	14 mm
Reach ins.	0.75	0.75	0.75	0.75
Reach mm	1.875	1.875	1.875	1.875
Gap ins.	0.025	0.025	0.025	0.025
Gap mm	0.625	0.625	0.625	0.625

Ignition timing

Crankshaft position (B.T.D.C.) F.A.	38°
Piston position (B.T.D.C.) F.A.	.357 ins.
Piston position (B.T.D.C.) F.A.	9.0678 mm
Advance range	
Contact breaker	12°
Crankshaft	24°

*(Replace with 6 volt 17M6 for transistorised ignition)

1 General description

1 The spark necessary to ignite the petrol/air mix in each combustion chamber is derived from a battery and coil used in conjunction with a contact breaker to determine the precise moment at which the spark will occur. As the points separate the circuit is broken and a high tension voltage is developed across the points of the spark plug which jumps the air gap and ignites the mixture. Each cylinder has its own ignition circuit hence the need for three ignition coils and a triple contact breaker assembly.

2 When the engine is running, current produced by the alternator is converted into direct current by the silicone bridge rectifier and used to charge the battery.

3 Alternator output does not correspond directly to engine r.p.m. and a zener diode is used to prevent voltage overload at high speeds.

4 It is possible to use machines with or without a battery if a capacitor is fitted as part of the 2MC capacitor ignition system.

2 Checking alternator output

1 Specialised test equipment of the multi-meter type is essential to check alternator output with any accuracy. If the owner possesses such an instrument and has the technical ability to use it without causing damage to his machine, his meter or himself, then money can be saved by the effort and the cash outlay.
2 Simple electrical checks and meter readings are given in Chapter 6, Section 2. Remember when checking alternator voltage to have the meter selected to read A.C. volts otherwise you will damage the meter movement.

3 Ignition coils - checking

1 An ignition coil is a sealed unit designed to give long service without need of attention.
2 The three coils are rubber mounted into the electrical platform beneath the twin seat.
3 A coil will fail if damage is caused to the outer casing. This can fracture the fine wire used for the secondary winding. The primary winding has a mere 280 - 372 turns of enamel covered wire compared to the secondary 19,000 turns of much finer wire.
4 A simple continuity check can indicate if a coil is faulty and a coil substitution which causes the fault to switch from one spark plug to another is simple to carry out.
5 If in any doubt about a coil then renew it. If fractured the fine secondary wire can fuse together to give output but only until the machine is vibrated when it will fail again.
6 Don't forget to check both low and high tension connections if a malfunction occurs. The high tension cables can be removed as individual cables for test and renewal. Use high tension cable in remaking them.

4 Removing and replacing the contact breaker

1 On machines fitted with the Lucas 7CA contact breaker system it is necessary to remove the unit from the right-hand cover for overhaul or renewal. The contact breaker is operated by the exhaust camshaft.
2 The unit is protected at the front by a circular cover and gasket and has an oil seal fitted at the rear. To remove proceed as follows:
3 Disconnect the battery feed, remove three screws and then the cover and its gasket.
4 The centre bolt can be removed but the auto advance mechanism is a taper fit into the end of the exhaust camshaft. It is necessary to use service tool D782 screwed into the bolt hole. The tool must be screwed in to release the auto advance cam from the locking taper.
5 Remove the service tool and then the three pillar bolts and plain washers from the contact breaker plate. Remove the breaker plate and then the auto advance mechanism.
6 Disconnect the contact breaker connections at the snap connectors behind the gearbox and withdraw the leads through the grommet in the timing case. Remove the contact breaker assembly to the bench. Servicing the points is covered in Section 5 of this Chapter.
7 It is a good idea to scribe your own positioning marks on the assembly before removal for aid in reassembly.
8 Reassembly is the reverse procedure of disassembly but it is necessary to adjust the contact breaker points gap and check the accuracy of the ignition timing. The final tightening up is carried out at the end of these procedures.
9 When a satisfactory setting is achieved for the three cylinders retighten the contact breaker bolts. Finally refit the cover and gasket.

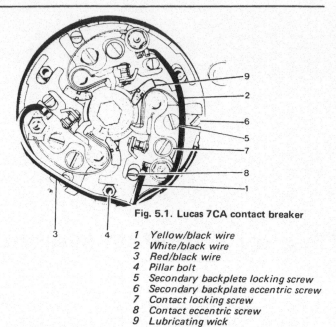

Fig. 5.1. Lucas 7CA contact breaker

1 Yellow/black wire
2 White/black wire
3 Red/black wire
4 Pillar bolt
5 Secondary backplate locking screw
6 Secondary backplate eccentric screw
7 Contact locking screw
8 Contact eccentric screw
9 Lubricating wick

5 Contact breaker points - removal, renovation and replacement

1 Faults in the ignition system can frequently be traced to the contact breaker points. The points can be incorrectly adjusted or neglected, leading to excessive wear, burning or pitting.
2 In the latter state the points must be renewed or removed for dressing. If it is necessary to remove a substantial amount of material before the faces can be restored, the points should be renewed.
3 If a fine emery cloth or carborundum stone is used for dressing the points they should be thoroughly cleaned afterwards with petrol. Try to obtain a slightly domed finish to the contacts. New contact breaker points are treated with a preservative which must be cleaned off before fitting.
4 In the event of the contacts needing a complete check, first remove the complete contact breaker unit from the machine.
5 To remove the moving contacts, unscrew the nut securing the low tension lead and remove the lead and nylon bush. Remove the spring and contact point from its pivot spindle.
6 To reassemble, the nylon bush is fitted through the low tension connection tab and the spring location eye. Replace every item in its original position.
7 The contact breaker cam and the moving contact pivot post should have a very light trace of grease.
8 The Lucas contact breaker assembly type 7CA has three lubricating felt wicks. These should receive two drops of engine oil every 3,000 miles.

6 Contact breaker points - adjustment

1 The contact breaker base plate should be assembled with the red and black cables rear-most.
2 The correct contact breaker gap is 0.015 inches with the contacts fully open. The kickstarter pedal should be used to turn the engine until the nylon heel of a set of points aligns with the scribe mark on the cam. The feeler gauge should require no forcing between the points but should have no clearance. To adjust the gap slacken the contact point locking screw and rotate the contact point eccentric screw.
3 Turn the engine to adjust the other two sets of points in turn.
4 It is sometimes found that there is a discrepancy between the points gaps of the 7CA unit when the scribe mark of the cam is aligned with the nylon heels. If the discrepancy is greater than 0.003 it is probably caused by cam run-out and can be cured by tapping the cam with a soft metal drift until it seats correctly.

Cases have also occurred where the edge of one of the secondary backplates has fouled the cam. Contact between the cam and the backplate can result in the automatic advance unit remaining in the permanently retarded position, so if run-out is evident, both of these two faults should be investigated and remedied.

7 Capacitors - removal and replacement

1   Capacitors for the Lucas 7CA contact breakers are located on the coil mounting plate beneath the twinseat. The three are mounted on a common bracket and have a rubber protection cover. They connect into the low tension circuit by means of Lucar connectors.
2   Before replacing a capacitor first disconnect the connectors, then the cover. The connectors are secured to the common bracket by a nut and shakeproof washer. Reassembly is the reversal of disassembly.
3   If the engine becomes difficult to start, or if misfiring occurs, it is probable that a condenser has failed. It is rare for all three condensers to fail simultaneously unless they have been damaged in an accident. Examine the contact breaker points whilst the engine is running to see whether arcing is taking place and, when the engine is stopped, examine the faces of the points.
4   It is not possible to check a condenser without the necessary test equipment. It is therefore best to fit a replacement condenser and observe the effect on engine performance especially in view of the low cost of replacement.

8   Ignition timing - checking and resetting

1   Remove the chromium-plated cover over the contact breaker assembly. It is secured by three screws.
2   Check the contact gaps in each three positions. They should be between 0.014 and 0.016 inches (0.35 - 0.40 mm).
3   Firing order on the Triumph Trident and BSA Rocket 3 models is 1, 3, 2.
4   Connecting leads are colour coded as follows:
    Number one cylinder   — White and Black (right-hand cylinder)
    Number two cylinder   — Red and Black (centre cylinder)
    Number three cylinder — Yellow and Black (left-hand cylinder)
5   Remove the blanking plug for the 38 degree b.t.d.c. locating hole. This is found slightly inboard of the timing cover.
6   Remove the two rocker covers and then the three spark plugs.
7   To permit the engine to be turned over with the rear wheel, select second gear.
8   Rotate the engine forwards until top dead centre is registered on Number one cylinder. Both valves will be closed and clearance shown on both tappets.
9   Screw the timing plunger and body (service tool D1858) into the crankcase and apply light finger pressure to the plunger. If the wheel is turned backwards slowly it will allow the plunger to locate in the hole drilled in the crankshaft web. This is now the 38 degree b.t.d.c. position. If the service tool is not available the same effect can be achieved by placing a small socket spanner in the hole and using the shank of a drill as the plunger.
10 Remove the auto advance central bolt and fit an oversized washer under it - this will lock the cam in the fully advanced position.
11 The auto advance should be refitted so that number one cylinder points are just open (0.0015 inches) when the auto advance is locked in the fully advanced (or fully clockwise) position.
12 This setting should be rechecked and if necessary readjust number one points (black and white cable connection) to achieve the correct gap. When satisfactory withdraw the D1858 service tool.
13 Establish the t.d.c. position on the compression stroke for Number three cylinder. Again rotate the engine backwards to allow the plunger tool to locate. The contact points should commence to break with the auto advance unit still locked in the fully

advanced position. If not, adjust the gap again on Number three position (yellow and black leads) contacts with the eccentric screw. Retighten the screw and withdraw the plunger. Repeat the procedure for Number two cylinder.
14 Finally remove the oversized washer under the auto advance unit centre bolt. On no account must the auto advance unit be disturbed from its position when retightening the bolt.
15 Remove the timing plunger and body and replace the blanking plug and fibre washer. Refit the spark plugs and rocker covers.

9   Ignition timing with a stroboscope

1   It is unlikely that the average machine owner will have access to an expensive item such as a stroboscope or perhaps knows what such an instrument is. However the method is given for the enthusiast or the keen hobbyist who wants perfection from his machine.
2   Some stroboscopes are operated from a twelve volt battery. In which case do not use the one on the machine but borrow one from an alternative source (car).
3   The stroboscope should then be connected to the battery and its probe to the right-hand spark plug.
4   The triangular patent plate next to the contact assembly should be suspended by the base screw of the three retaining it. There are three scribed lines on the rotor at 120° intervals.
5   Remove the contact breaker cover plate and gasket. The points gap should have been accurately set prior to this exercise.
6   Start the engine and use the base screw as a pointer direct the strobe beam at the pointer and the rotor mark. At 2,000 r.p.m. or more the pointer and the line should coincide. If this does not occur readjust the points for this cylinder (black/white).
7   Repeat the exercise for the centre cylinder (red/black lead points) and the left-hand cylinder (yellow/black lead points).
8   When satisfactory timing is achieved switch off the engine, ensure all the contact breaker screws are tight. Replace the gasket and cover and refit the patent plate. After removing the stroboscope, ensure the spark plug caps are securely fitted.

10 Automatic advance unit - removal, examination and replacement

1   Fixed ignition timing is of little advantage as the engine speed increases and it is therefore necessary to incorporate a method of advancing the timing by centrifugal means. A balance weight assembly located behind the contact breaker, linked to the contact breaker cam is employed in the case of the Triumph unit-construction Trident. It is secured to the exhaust camshaft by a bolt that passes through the centre of the contact breaker cam. It can be withdrawn without need to remove the timing cover, if the contact breaker assembly is removed first.
2   When the assembly is removed from the machine, it is advisable to make a note of the degree figure stamped on the back of the cam unit. This relates to the ignition advance range and, as the previous section has indicated, must be known for accurate static timing.
3   The unit is most likely to malfunction as the result of condensation, which will cause rusting to take place. This will immediately be evident when the assembly is removed. Check that the balance weights move quite freely and that the return springs are in good order. Before replacing the assembly by reversing the dismantling procedure, lubricate the balance weight pivot pins and the cam spindle, and place a light smear of grease on the face of the contact breaker cam. Lubricate the felt pad that bears on the contact breaker cam.

11 Capacitor ignition - alternative system

1   On some models the model 2MC system of capacitor ignition is fitted or can be fitted.
2   This is mainly used for competition riding and enables the

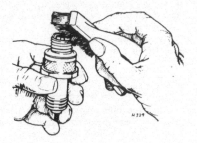

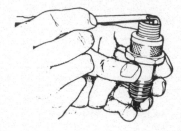

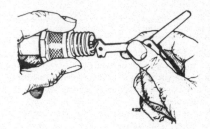

*Cleaning deposits from elec-
trodes and surrounding area
using a fine wire brush*

*Checking plug gap with feeler
gauges*

*Altering the plug gap. Note use
of correct tool*

Fig. 5.2a. Spark plug adjustment

*White deposits and damaged
porcelain insulation indicating
overheating*

*Broken porcelain insulation
due to bent central electrode*

*Electrodes burnt away due to
wrong heat value or chronic
pre-ignition (pinking)*

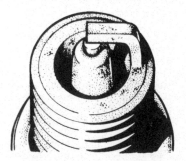

*Excessive black deposits
caused by over-rich mixture
or wrong heat value*

*Mild white deposits and elec-
trode burnt indicating too
weak a fuel mixture*

*Plug in sound condition with
light greyish brown deposits*

Fig. 5.2b. Spark plug electrode conditions

12.1 The Boyer transistorised ignition system

12.5 Magnetic rotor replaces the contact breaker cam

rider to use the machine with or without a battery fitted. It is invaluable for emergency use in battery failure.

3   A machine so equipped can be started without a battery being fitted and the lighting circuit will function as normal. The voltage ripple found on the D.C. is also reduced.

4   In a stationery position however the parking lights cannot operate without a battery.

5   To obtain this facility it is necessary to fit to the machine a high capacity spring mounted electrolytic condenser type 2MC.

6   The electrolytic capacity will store up a large charge across its plates to ensure that sufficient current flows through the ignition coils at the moment of contact opening, and produces the requisite ignition spark.

7   Fit the capacitor to the instructions given in the kit CP.210 and follow the connecting and storing procedures. The writer's experience is that electrolyte condensers need handling with great respect but the following is a fairly easy way of checking one.

8   Connect the capacitor to a 12 volt battery (correctly connected) for 5 seconds. Disconnect and after 5 minutes check the voltage. It should not be less than 9 volts. Below this voltage calls for a change of capacitor as the unit is leaking. It is wise to ensure that the battery is 12 volts before testing. If a voltmeter is not available a good spark should be obtained across the shorted terminals after the charging has been carried out.

12 Transistorised ignition - general

1   Some Triumph Trident and BSA Rocket 3 models are equipped with transistorised ignition. The demonstration model for this manual was equipped with the Boyer system.

2   Makers of transistorised kits say that the weak point on three cylinder models is the contact breaker advance and retard assembly. It is claimed that the high drag of three contact breaker heels on the single cam tends to cause retardation at high revolutions. A criticism of the normal system is of the difficulty experienced in obtaining accurate timing on each cylinder with three separate sets of points. In addition cam heel wear and contact wear affects the timing. There is nothing to wear on the transistorised system so in the long run maintenance costs are reduced. The initial cost of the kit is £30 + VAT at December 1973 levels and it is necessary to purchase three new 6 volt H.T. coils at a cost of about £9. These are Lucas type 17M6 or equivalent and are to replace the standard 12 volt ignition coils.

3   Transistorised ignition kits such as the Boyer Mark 2A are supplied complete with detailed fitting instructions that must be studied carefully before fitment.

4   Timing should be set up by stroboscope and this item is one

that only the auto electric specialist will possess. The ignition unit sits in the area vacated by the contact breaker assembly and adjustment is achieved by moving the unit in its elongated slots. Let the dealer carry out the final adjustment with his instruments.

5   On transistorised systems the auto advance system is removed and replaced by a magnetic rotor. The demonstration model was equipped with such. It is held by a short A.F. bolt.

13 Spark plugs - checking and resetting the gap

1   A 14 mm spark plug is fitted to each of the three cylinders of the Triumph Trident and BSA Rocket 3 machines.

2   The gaps for these plugs should be set at 0.025 inches which is 0.625 mm. It is possible that for racing or extensive use in hot climates other types of plug may be necessary, in which case follow the plug manufacturer's instructions.

3   The full life for the spark plugs is 12,000 miles or 20,000 km. They should be removed, cleaned and the gaps reset every 3,000 miles or 4,800 km.

4   Remove the plugs with a correct size box spanner, 13/16 inches or 19.5 mm. If difficulty is experienced in the removal, spread a small amount of penetrating oil around the base rear and the thread and allow a short time for penetration before recommencing the removal.

5   Identify each plug by its cylinder number. This will assist if fault diagnosis is necessary.

6   The chart in this section showing spark plug electrode conditions should assist in such fault diagnosis.

7   To reset the gap, bend the outer electrode to bring it closer to the inner electrode and ensure that a 0.025 inch feeler gauge can just be inserted. Bending of the centre electrode can cause it to crack and eventually disintegrate. If the broken particles fall into the engine subsequent damage can occur.

8   Carry a set of spare spark plugs. They may get the machine back home in the event of a malfunction.

9   Overtightening a spark plug can cause stripping of the cylinder head thread particularly the light alloy type. It will then be necessary for your dealer to fit a Helicoil insert to get the machine back onto the road.

10 Clean deposits from the electrodes and surrounding area with a wire brush and then with a proprietary plug cleaner (with petrol as a suitable alternative). Smear a small amount of graphite grease on the threads, which should previously have been burnished with the wire brush.

11 Make sure the plug caps are a good fit and free from cracks. The caps contain the suppressors that eliminate radio or T.V. interference.

14 Fault diagnosis: ignition system

Sympton	Cause	Remedy
Engine will not start	No spark at plug	Check whether contact breaker points are opening and also whether they are clean. Check wiring for break or short circuit.
Engine fires on one cylinder	No spark plug or defective cylinder	Check as above, then test ignition coil. If no spark, see whether points arc when separated. If so, renew condenser.
Engine starts but lacks power	Automatic advance unit stuck or damaged	Check unit for freedom of action and broken springs.
	Ignition timing retarded	Verify accuracy of timing. Check whether points gaps have closed.
Engine starts but runs erratically	Ignition timing too far advanced	Verify accuracy of timing. Points gaps too great.
	Spark plugs too hard	Fit lower grade of plugs and re-test.

Chapter 6 Frame and forks

Contents

Specifications

Overall dimensions

Length ins	86
Length cms	218
Width ins	33
Width cms	84
Seat height ins	32
Seat height cms	81
Weight lbs	460
Weight kg	209
Ground clearance ins	6.5
Ground clearance cms	16

Head races

(Top and Bottom) Number of balls	20
Ball diameter ins	0.25
Ball diameter mm	6.35

Rear suspension

Type	Swinging fork controlled by combined coil spring/Hydraulic damper units
Extended distance between ins	12.875
Centre mm	32.66
Compressed distance between ins	10.375
Centre mm	23.36

Front forks

Type	Telescopic oil damped
Spring details	
Free length inches	9.688 - 9.812
Free length mm	246.075 - 249.225
Number of working coils	15½
Spring rate lb-ins	32.5
Spring rate kg-m	4.485
Gauge s.w.g.	5.0
Damper sleeve	
Length inches	2.125
Length mm	53.975

Internal diameter inches	1.387 - 1.393
Internal diameter mm	35.2298 - 35.3822

Bush details

Material	Sintered bronze
Top bush length inches	1.0
Top bush length mm	25.4
Top bush outer diameter inches	1.498 - 1.499
Top bush outer diameter mm	38.0492 - 38.0746
Top bush inner diameter inches	1.3065 - 1.3075
Top bush inner diameter mm	33.185 - 33.2105
Bottom bush length inches	0.870 - 0.875
Bottom bush length mm	22.098 - 22.225
Bottom bush outer diameter inches	1.4935 - 1.4945
Bottom bush outer diameter mm	37.945 - 37.960
Bottom bush inner diameter inches	1.2485 - 1.2495
Bottom bush inner diameter mm	31.712 - 31.7373
Stanchion diameter inches	1.3025 - 1.3030
Stanchion diameter mm	33.0835 - 33.0962
Working clearance in top bush inches	0.0035 - 0.0050
Working clearance in top bush mm	0.0889 - 0.127
Fork leg bore diameter inches(sliding tube)	1.498 - 1.500
Fork leg bore diameter mm (sliding tube)	38.049 - 38.10
Working clearance in bottom bush ins	0.0035 - 0.0065
Working clearance in bottom bush mm	0.0889 - 0.165

Swinging fork

Bush type	Pre-sized steel backed Glacier WB 1624
Bush bore diameter inches	1.4460 - 1.4470
Bush bore diameter mm	36.7284 - 36.7538
Sleeve diameter inches	1.4445 - 1.4450
Sleeve diameter mm	36.693 - 36.702
Distance between fork ends ins	7.5
Distance between fork ends mm	190.5
Housing diameter ins	1.1250 - 1.1262
Housing diameter mm	31.75 - 32.00
Spindle diameter ins	0.810 - 0.811
Spindle diameter mm	20.574 - 20.599
Inner spacer tube diameter inches	0.812 - 0.817
Inner spacer tube diameter mm	20.624 - 20.651
Outer spacer tube diameter inches	0.9972 - 0.9984
Outer spacer tube diameter mm	25.328 - 25.359
Spacer tube clearance inches (swinging arm spindle)	0.001 - 0.007
Spacer tube clearance mm (swinging arm spindle)	0.0254 - 0.1778
Clearance in bush inches	0.0016 - 0.0040
Clearance in bush mm	0.0306 - 0.1016
Inner bush bore diameter inches	1.0 - 1.0012
Inner bush bore diameter mm	25.4 - 25.43
Outer bush bore diameter inches	1.125
Outer bush bore diameter mm	28.5

1 General description

A full cradle frame is fitted to the Triumph and BSA three cylinder models, in which the front down tube branches into two duplex tubes at the lower end which form the cradle for the unit-construction engine/gear unit.

Rear suspension is provided by a swinging arm assembly that pivots from a lug welded to the vertical tube immediately to the rear of the gearbox. Movement is controlled by two hydraulically-damped rear suspension units, one on each side of the subframe. The units have three-rate adjustment, so that the spring loading can be varied to match the conditions under which the machine is to be used.

Front suspension is provided by telescopic forks of conventional design.

2 Front forks - removal from frame

1   The front fork assembly is telescopic in operation. It uses steel stanchions ground over the sliding portion of their length. The bottom members slide in sintered bronze bushes. The lower bushes are secured to the bottom of each stanchion and the top ones being shouldered to rest on the top of the fork bottom members. They are secured at the bottom by chromium plated dust excluder sleeve nuts.

2   External main springs are supported at the base by dust excluder sleeve nuts. These contain garter type oil seals. Oil is used both for lubrication and damping. The stanchions have drilled bleed holes. Shuttle valves are fitted to the lower end of each stanchion and retained by circlips.

3   Secured inside the bottom members by hexagon headed bolts and recessed into the spindle cutaway are fitted cone shaped restrictors.

4   It is unlikely that the front forks will need to be removed from the frame as a complete unit, unless the steering bearings require attention or the forks have been damaged.

5   Commence operations by draining the fork legs into a suitable tray by unscrewing the two small drain plugs at the base of the legs and pumping the legs up and down several times to assist drainage.

6 It is advisable to detach the forks as a complete unit and then remove the top lug only when the **stanchions and middle lug** assembly is lower than the frame.

7 The front wheel must be raised about six inches clear of the ground by supporting the engine on a stand.

8 Remove the following as detailed in the appropriate Chapters and Sections:

Front wheel (On disc brake models disconnect the hydraulic connections).

Front mudguard (Remove the stay nuts).

Headlamp unit - detach only.

Throttle cable - detach only.

Air control cable (some models) - detach only.

9 Remove the handlebars by unscrewing the two self locking nuts securing the eye bolts beneath the upper yoke. Metalistic bushes are fitted into the fork top lug. The hemispherical washers must be fitted when reassembling with the rounded side towards the head lug.

10 Remove the steering damper knob, slacken the upper yoke pinch bolt, unscrew the sleeve nuts. An Allen key will be needed for the exercise.

11 Remove the top cap nuts from the stanchions.

12 Disconnect the Zener diode connector and also the earthing wire from the mounting bolts.

13 The headlamp, nacelle and fork shrouds can remain in position if the fork pinch bolts are removed.

14 If the fork is supported give the top lug a tap underneath to release it from the stanchion locking tapers.

15 Lower the stanchions and the lower yoke assembly from the steering head.

16 The top ball race can be left in situ if carefully handled. The lower race balls can be collected during the overhaul.

17 It is possible to remove the stanchions without disturbing the head races and middle and top lugs. This requires a drift, BSA/ Triumph number 61-3824 which is essential for this operation.

18 Remove the top nuts, slacken off the lower yoke pinch bolts and screw in the special drift which drives the stanchion downwards and out. The use of a punch or drift is prohibited as it will damage or distort the internal threads.

---

3 Front forks - dismantling and examining the fork legs

---

1 If the springs only require to be changed all that is required is to remove the rubber gaiters and pull out the old springs. Liberally grease and then refit the new springs. It is necessary on some models to remove a cap nut.

2 To remove the stanchions, screw Triumph/BSA service tool **61-3824** into the top of the stanchion. An old cap nut can be used if the correct tool cannot be obtained. Drive the stanchions out of the lug and collect the spring abutments, springs, gaiters and clips (4 off BSA models).

3 Remove the fork top shrouds.

4 The chrome dust excluder-cum-sleeve nut has holes for locating service tool 61-6017 (BSA and Triumph). The nut has a right-hand thread. Give the spanner a light tap with a soft hide hammer to start the movement. The nut had no locating holes on the demonstration model.

5 When the dust excluder nut is away a few sharp pulls should release the stanchion, bush and damper valve assembly from the bottom member. A damper valve nut required removal on the demonstration model.

6 To remove the restrictor from the bottom member, release the securing bolt. There is some variation in the damper tube assemblies as between BSA and Triumph models and years of manufacture. Take the assembly with you to the dealer if you require an identical replacement.

7 The hexagon headed oil restricter securing bolt which is counterbored into the wheel spindle lug is sealed by means of an aluminium washer which should be withdrawn from the counterbore when the bolt is removed and carefully stowed.

8 The shuttle valves are retained in the bottom end of each

2.8 Remove the front mudguard stay nuts

2.9 An Allen key will be needed to release the top yoke clamps

3.1A Fork legs will have to be removed if springs need to be changed

3.1B It is necessary on some models to remove a cap nut

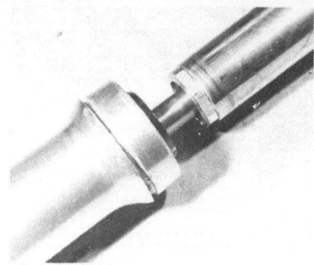

3.5A Removal of dust excluder reveals nut

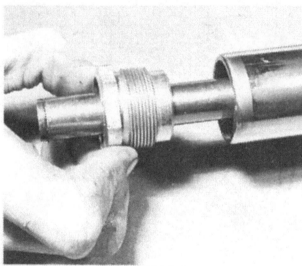

3.5B To free damper valve, remove nut

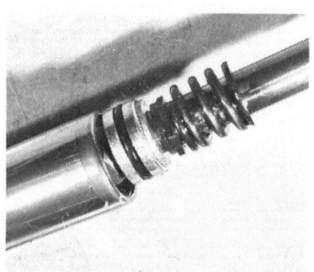

3.6 Damper design may not be identical

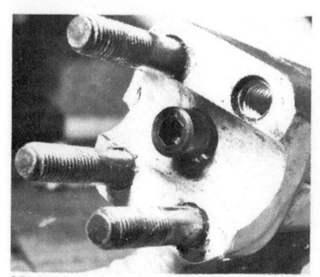

3.7A Allen key retains oil restrictor assembly

3.7B The complete damper assembly

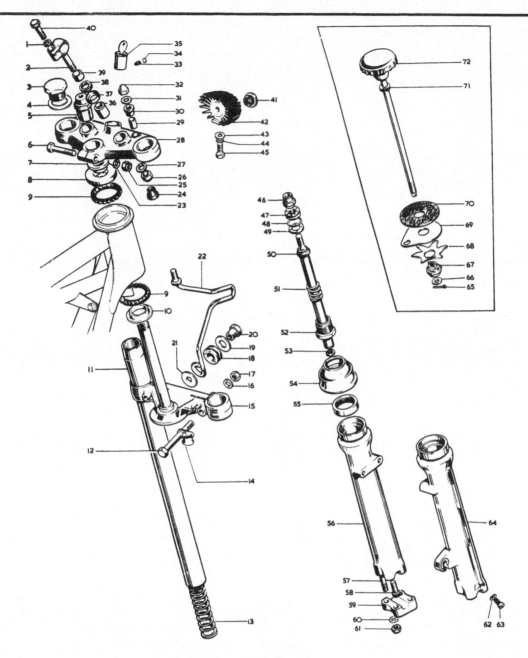

Fig. 6.1. Front fork (Rocket 3)

1	Washer - 2 off	19	Washer - 2 off	37	Handlebar mounting cup - 2 off	55	Oil seal - 2 off
2	Clamp - 2 off	20	Sleeve nut - 2 off			56	Lower left-hand fork leg
3	Nut - 2 off	21	Washer - 2 off	38	Handlebar mounting rubber - 2 off	57	Stud - 8 off
4	Washer - 2 off	22	Headlamp mounting bracket			58	Cap screw - 2 off
5	Nut	23	Washer	39	Handlebar mounting distance piece - 2 off	59	Wheel spindle cap - 2 off
6	Pinch bolt	24	Bush - 2 off			60	Washer - 8 off
7	Collar	25	Nut	40	Handlebar clamp bolt - 2 off	61	Nut - 8 off
8	Top cone and dust cover	26	Nut - 2 off	41	Heat sink plug	62	Fibre washer - 2 off
9	Ball race - 2 off	27	Washer - 2 off	42	Diode heat sink	63	Drain plug - 2 off
10	Cone	28	Top yoke with lock and keys	43	Washer	64	Lower right-hand fork leg
11	Stanchion - 2 off	29	Distance piece - 2 off	44	Washer	65	Split pin
12	Pinch bolt - 2 off	30	Bush - 2 off	45	Bolt	66	Washer
13	Fork spring - 2 off	31	Washer - 2 off	46	Damper valve nut - 2 off	67	Steering damper nut
14	Clip	32	Nut - 2 off	47	Damper valve - 2 off	68	Spring
15	Bottom yoke	33	Grub-screw	48	'O' ring - 2 off	69	Damper anchor plate
16	Washer - 2 off	34	Sealing washer	49	Washer - 2 off	70	Friction disc
17	Nut - 2 off	35	Steering lock complete with keys	50	Damper tube - 2 off	71	Spring washer
18	Grommet - 2 off	36	Handlebar clamp bush - 2 off	51	Recoil spring - 2 off	72	Steering damper knob and rod
				52	End plug - 2 off		
				53	Cap screw seal - 2 off		
				54	Dust seal - 2 off		

Fig. 6.2. Front fork (Trident)

1 Fork assembly
2 Top yoke
3 Pinch bolt
4 Nut
5 Washer
6 Cap screw - 2 off
7 Lock c/w 2 keys
8 Key
9 Grub screw
10 Sealing washer
11 Bottom cone - 2 off
12 Bottom yoke and stem
13 Pinch bolt
14 Washer - 4 off
15 Nut - 2 off
16 Stanchion - 2 off
17 End plug - 2 off
18 Main spring - 2 off
19 Outer member - LH (assembly)
20 Outer member - RH (assembly)
21 Outer member LH
22 Outer member RH
23 Wheel spindle cap
24 Wheel spindle cap
25 Stud - 8 off
26 Nut - 8 off
27 Washer - 8 off
28 Oil seal - 2 off
29 Stud - 2 off
30 Drain plug - 2 off
31 Washer - 2 off
32 'O' ring - 2 off
33 Recoil spring - 2 off
34 Cap screw - 2 off
35 Cap screw seal - 2 off
36 Damper tube and valve assembly - 2 off
37 Top cap nut - 2 off
38 Washer - 2 off
39 Steering stem nut
40 Cap screw - 2 off
41 Rubber ring - 2 off
42 Scraper sleeve - 2 off
43 Outer cover - LH
44 Outer cover - RH
45 Headlamp bracket - LH
46 Headlamp bracket - RH
47 Rubber mounting - 2 off
48 Rubber buffer - 4 off
49 Backing washer - 4 off
50 Attachment bolt - 4 off
51 Nut

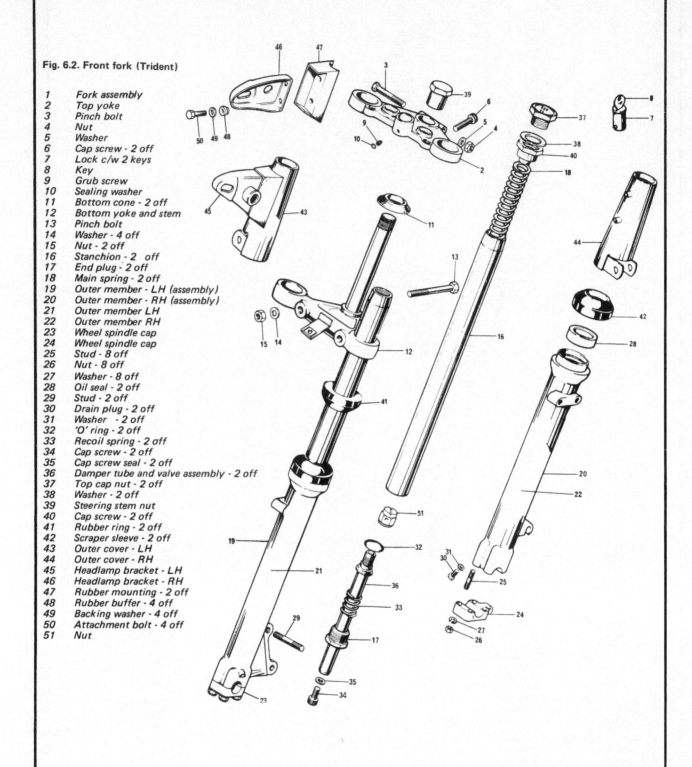

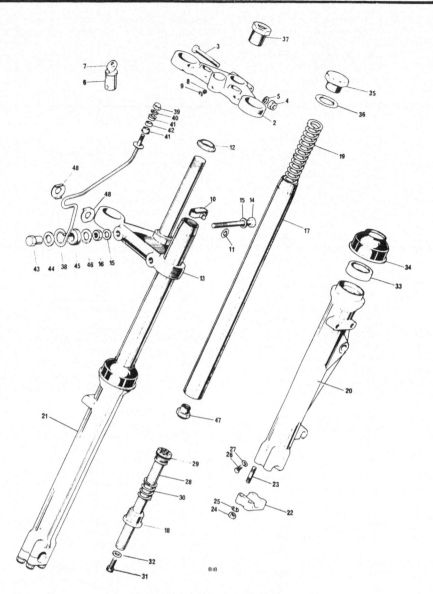

Fig. 6.3. Front fork (Trident disc brake models)

1	Fork assembly	25	Washer - 8 off
2	Top yoke	26	Drain plug - 2 off
3	Pinch bolt	27	Washer - 2 off
4	Nut	28	Damper tube and valve assembly - 2 off
5	Washer	29	'O' ring - damper valve - 2 off
6	Lock c/w 2 keys	30	Recoil spring - 2 off
7	Key	31	Cap screw - 2 off
8	Grub screw	32	Cap screw seal - 2 off
9	Sealing washer	33	Oil seal - 2 off
10	Brake cable retainer	34	Scraper sleeve outer member - 2 off
11	Starlock washer	35	Top cap nut - 2 off
12	Bottom cone	36	Washer - 2 off
13	Bottom yoke and stem	37	Steering stem nut
14	Pinch bolt - 2 off	38	Headlamp bracket - LH
15	Washer - 4 off	38	Headlamp bracket - RH
16	Nut - 2 off	39	Nut) - 2 off
17	Stanchion - 2 off	40	Washer) Headlamp bracket, top fixing - 2 off
18	End plug - 2 off	41	Bush) - 4 off
19	Main spring - 2 off	42	Spacer) - 2 off
20	Outer member - RH	43	Sleeve nut) - 2 off
21	Outer member - LH	44	Washer) Headlamp bracket, bottom fixing - 2 off
22	Wheel spindle cap - 2 off	45	Grommet) - 2 off
23	Stud - 8 off	46	Washer) - 2 off
24	Nut - 8 off	47	Nut - 2 off
		48	Clamp washer - 4 off

stanchion by the bottom bearing retaining nut. Circlips are fitted to prevent the damper valves recessing into the stanchions.

9   Later damper units fitted to Triumph models have a shuttle valve damper attached to the lower end of the fork stanchions. The damper is retained by a sleeve nut which also holds the bottom bearing of the fork leg. A circlip or spring in front of this nut prevents the valve from passing into the stanchion. This type of damper assembly is easily recognised by the eight holes for oil bleeding above the bottom bearing location.

10 The parts most liable to become damaged in an accident are the fork stanchions, which will bend on heavy impact. To check for misalignment, roll the stanchion on a sheet of plate glass, when any irregularity will be obvious immediately. It is possible to straighten a stanchion that has bowed not more than 5/32 in. out of true but it is debatable whether this action is desirable. Accident damage often overstresses a component and because it is not possible to determine whether the part being examined has suffered in this way, it would seem prudent to renew rather than repair.

11 Check the top and bottom fork yokes which may also twist or distort in the event of an accident. The top yoke can be checked by temporarily replacing the stanchions and checking whether they lay parallel to one another. Check the lower fork yoke in the same manner, this time with the stanchions inserted until about 6½ inches protrude. Tighten the pinch bolts before checking whether the stanchions are parallel with one another. The lower yoke is made of a malleable material and can be straightened without difficulty or undue risk of fracture.

12 It is possible for the lower fork legs to twist and this can be checked by inserting a dummy wheel spindle made from 11/16 inch diameter bar and replacing the split retaining clamps. If a set square is used to check whether the fork leg is perpendicular to the wheel spindle, any error is readily detected. Renewal of the lower fork leg is necessary if the check shows misalignment.

13 The top and bottom bushes can be measured and the results compared with those given at the front of this Chapter in the Specifications Section. These measurements are for the modified type made of sintered bronze. The bushes can also be checked against their respective mating surfaces. Put the top bush over the stanchion and at about eight inches from the bottom of the stanchion check the diametrical clearance at the bush. Too much play indicates the need for renewal. The bottom bush must be fitted to the stanchion and inserting the stanchion into the bottom member to a depth of about eight inches, whilst the diametrical clearance is estimated from the amount of "play".

14 Examine the main springs for fatigue and cracks and check that both are of equal length and also within ¼ inch of the original length. Check against the data given in the Specifications Section at the beginning of this Chapter.

15 Before replacing the main springs, ensure the outer member oil seals are undamaged.

4   Front forks - examination of the steering head races

1   If the steering head races have been dismantled, it is advisable to examine them prior to reassembling the forks. Wear is usually evident in the form of indentations in the hardened cups and cones, around the ball track. Check that the cups are a tight fit in the steering column head lug.

2   If it is necessary to renew the cups and cones, use a drift to' displace the cups by locating with their inner edge. Before inserting the replacements, clean the bore of the head lug. The replacement cups should be drifted into position with a soft metal drift or even a wooden block. To prevent misalignment, make sure that the cups enter the head lug bore squarely. The lower cone can be levered off the bottom fork yoke with tyre levers; the upper cone is within the top fork yoke and can be drifted out. Clean up any burrs before the new replacements are fitted. A length of tubing which will fit over the head stem can be used to drive the lower cone into position so that it seats squarely.

3   When the cups and cones are replaced, discard the original ball bearings and fit a new set. It is false economy to re-use the originals in view of the very low renewal cost. Forty ¼ inch dia-

meter balls are required, 20 for each race. Note that when the bearing is assembled, the race is not completely full. There should always be space for one bearing, to prevent the bearings from skidding on one another and wearing more rapidly. Use thick grease to retain the ball bearings in place whilst the forks are being offered up.

5   Front forks - reassembling the fork legs

1   The fork legs are reassembled by following the dismantling procedure in reverse. Make sure all of the moving parts are lubricated before they are assembled. Fit new oil seals, regardless of the condition of the originals.

2   Before refitting the fork stanchions, make sure the external surfaces are clean and free from rust. This will make fitting into the fork yokes at a later stage much easier. Oil, or lightly grease, the outer surfaces after removing all traces of the emery cloth or other cleaner used.

3   When refitting the fork gaiters (if originally fitted) check that they are positioned correctly. The small hole near the area where the clip fastener is located should be at the bottom.

6   Front forks - refitting to frame

1   If it has been necessary to remove the fork assembly complete from the frame, refitting is accomplished by following the dismantling procedure in reverse. Check that none of the ball bearings are displaced whilst the steering head stem is passed through the headlug; it has been known for a displaced ball to fall into the headlug and wear a deep groove around the headstem of the lower fork yoke.

2   Take particular care when adjusting the steering head bearings. The blind or sleeve nut should be tightened sufficiently to remove all play from the steering head bearings and no more. Check for play by pulling and pushing on the fork ends and make sure the handlebars swing easily when given a light tap on one end.

3   It is possible to overtighten the steering head bearings and place a loading of several tons on them, whilst the handlebars appear to turn without difficulty. On the road, overtight head bearings cause the steering to develop a slow roll at low speeds.

4   Before the plated top fork nuts are replaced, do not omit to replace the drain plug in each fork leg and to refill each leg with the correct quantity of SAE 20 oil. Each leg holds 200 cc - see Specifications).

5   Difficulty will be experienced in raising the fork stanchions so that their end taper engages with the taper inside the top fork. Triumph/BSA service tool 61-3824 is specified for this purpose; if the service tool is not available, a wooden broom handle screwed into the inner threads of the fork stanchion can often be used to good effect.

6   Before final tightening, bounce the forks several times so that the various components will bed down into their normal working locations. This same procedure can be used if the forks are twisted, but not damaged, as the result of an accident. Always retighten working from the bottom upward.

7   Frame assembly - examination and renovation

1   If the machine is stripped for a complete overhaul, this affords a good opportunity to inspect the frame for cracks or other damage which may have occurred in service. Check the front down tube immediately below the steering head and the top tube immediately behind the steering head, the two points where fractures are most likely to occur. The straightness of the tubes concerned will show whether the machine has been involved in a previous accident.

2   If the frame is broken or bent, professional attention is required. Repairs of this nature should be entrusted to a competent repair specialist, who will have available all the necessary jigs and mandrels to preserve correct alignment. Repair work of this

3.15 Check to ensure the oil seals are undamaged

6.4 Plated nuts pass through instrument mountings

8.6 Remove the two bolts securing suspension unit to swinging arm

nature can prove expensive and it is always worthwhile checking whether a good replacement frame of identical type can be obtained from a breaker or through the manufacturer's Service Exchange Scheme. The latter course of action is preferable because there is no safe means of assessing on the spot whether a secondhand frame is accident damaged too.

3   The part most likely to wear during service is the pivot and bush assembly of the swinging arm rear fork. Wear can be detected by pulling and pushing the fork sideways, when any play will immediately be evident because it is greatly magnified at the fork end. A worn pivot bearing will give the machine imprecise handling qualities which will be most noticeable when traversing uneven surfaces.

8   Swinging arm - examination and renovation

1   If the swinging arm is to be removed with the engine still in the frame proceed as follows:

2   Remove the side panels.

3   Disconnect the chain after removing the spring link. Remove the front anchor stay securing bolt from the rear brake torque arm. Unscrew the brake operating rod adjuster nut and remove the speedometer cable from the rear wheel drive. Slacken the wheel spindle nuts and take off the wheel.

4   Remove the large nut from the right-hand engine plate. Remove the two long, two short and the large central bolt from the left-hand engine plate. After disconnecting the stop lamp leads, remove the plates.

5   Slacken off the rear chain guard bolt and remove the front chain guard securing bolt.

6   Remove the two bolts which secure the suspension units to the swinging arm and disconnect the rear chain oiler pipe.

7   To withdraw the swinging arm spindle on Triumph models, bend back the tab washer and unscrew the locking nut from the left-hand side of the spindle. The spindle should be unscrewed until it is free to be withdrawn. This releases the swinging arm which can then be removed and the lipped end plates with 'O' rings, outer sleeves and distance tubes removed. With the swinging arm out of the machine check that it is straight and not twisted.

8   To withdraw the swinging arm spindle on BSA models remove the 13/16 inch U.N.F. nut and star washer from the right-hand side of the spindle and unscrew the nut and bolt from the location plate on the left-hand end of the spindle. The spindle is withdrawn from the left-hand side of the machine. If corrosion exists then a drift of nor more than 0.805 inch diameter can be used to drive out the spindle.

9   All parts of the swinging arm should be thoroughly cleansed in paraffin and inspected for wear. The make up of the swinging arm for the BSA and Triumph differs in some respects but on the Triumph model particular attention should be paid to the fit of the two outer sleeves in the swinging arm bushes. On either model remove and replace if wear exists. A soft metal drift can be used to displace the bushes which are a good press fit in the pivot tube.

10 All parts should be reassembled in the order of removal. The spindle and distance tube should be fully greased. This also applies to the sleeves and bushes.

11 On Triumph models the 'O' rings should be inserted into the lipped end plates and pushed over the ends of the swinging arm cross tube whilst offering the arm to the pivot lug and inserting the bolt from the right-hand side. The bolt should be tightened until the arm can just be moved up and down with little effort. Fit the lock nut and tab washer and tighten the nut.

12 If there is side play and the bushes are within specification, the distance tube can be reduced in length by careful filing.

13 If new bushes are to be fitted they are normally the steel backed type and if pressed in carefully using a smear of grease to aid assembly, they will give the correct working clearance without need for reaming. The bushes can be fitted by using a suitable sized soft drift and hammer. The bush must enter squarely and no burrs should be permitted.

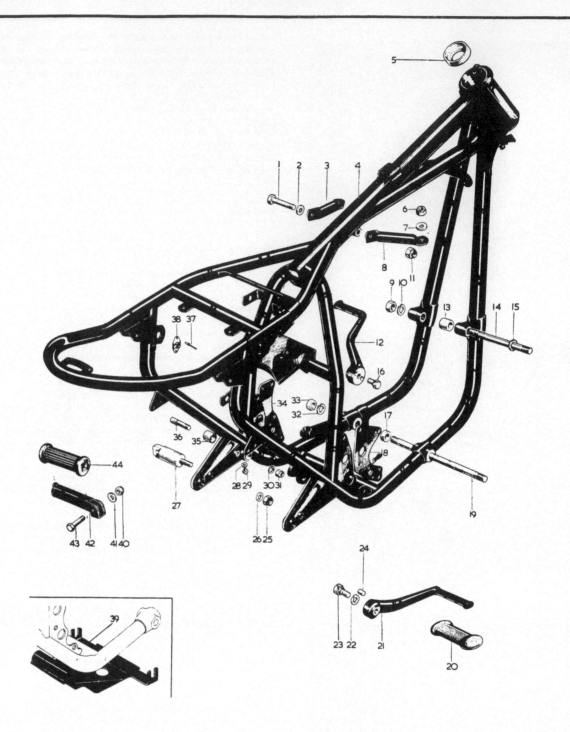

Fig. 6.4. Frame assembly (Rocket 3)

1 Bolt	12 Footrest - left	24 Footrest peg - 4 off	35 Brake pedal bush
2 Washer	13 Distance piece	25 Locknut - 2 off	36 Brake pedal stop peg
3 Steady stay	14 Stud	26 Washer - 2 off	37 Oddie clip pin
4 Frame complete	15 Washer - 2 off	27 Anchor bolt - 2 off	38 Oddie clip
5 Steering head cup - 2 off	16 Bolt	28 Fibre washer - 2 off	39 Engine shield
6 Nut - 2 off	17 Distance piece	29 Grease nipple - 2 off	40 Nut - 2 off
7 Washer - 2 off	18 Engine plate	30 Spring washer	41 Washer - 2 off
8 Steady stay	19 Stud	31 Stop peg nut	42 Pillion footrest - 2 off
9 Nut - 2 off	20 Rubber	32 Spring washer	43 Pillion footrest bolt - 2 off
10 Washer - 2 off	21 Footrest - right	33 Nut	44 Pillion footrest rubber - 2 off
11 Nut	22 Spring washer - 2 off	34 Left-hand rear engine plate	
	23 Bolt		

8.7A Bend back the tab washer, remove nut ...

8.7B ... and unscrew spindle whilst pulling outwards

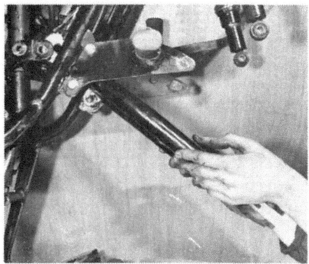
8.7C This releases the swinging arm

8.9A Sleeves must be a good fit

8.9B Note end caps have internal 'O' ring

8.10 The spindle passes through this central distance piece

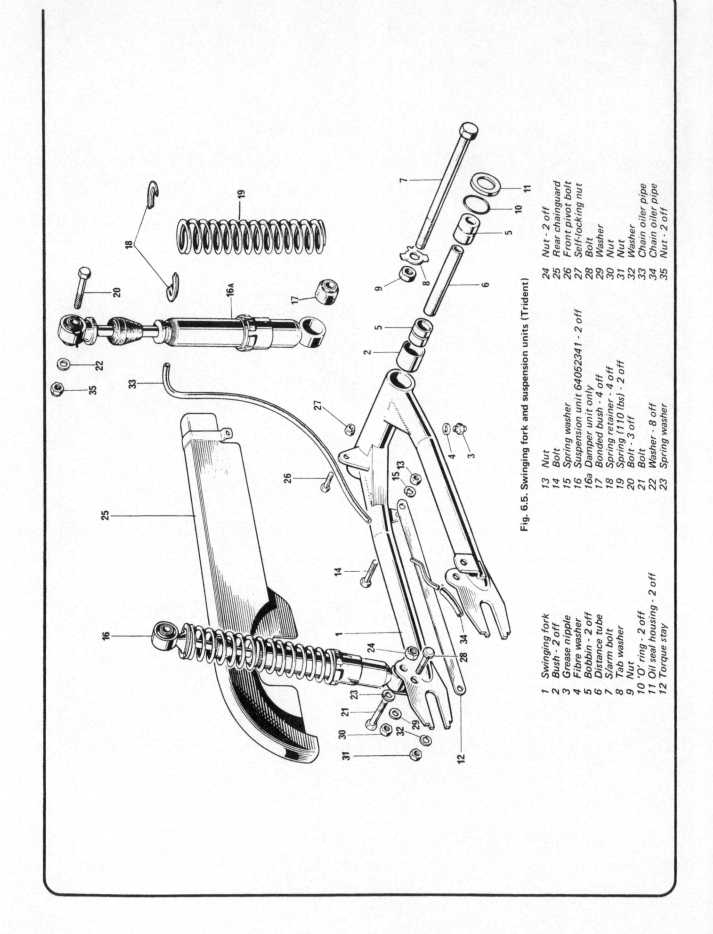

Fig. 6.5. Swinging fork and suspension units (Trident)

1   Swinging fork
2   Bush - 2 off
3   Grease nipple
4   Fibre washer
5   Bobbin - 2 off
6   Distance tube
7   S/arm bolt
8   Tab washer
9   Nut
10 'O' ring - 2 off
11 Oil seal housing - 2 off
12 Torque stay
13 Nut
14 Bolt
15 Spring washer
16 Suspension unit 64052341 - 2 off
16a Damper unit only
17 Bonded bush - 4 off
18 Spring retainer - 4 off
19 Spring (110 lbs) - 2 off
20 Bolt - 3 off
21 Bolt
22 Washer - 8 off
23 Spring washer
24 Nut - 2 off
25 Rear chainguard
26 Front pivot bolt
27 Self-locking nut
28 Bolt
29 Washer
30 Nut
31 Nut
32 Washer
33 Chain oiler pipe
34 Chain oiler pipe
35 Nut - 2 off

14 On models with bronze bushes the swinging fork pivot should be lubricated with a high pressure grease gun at least every 1,000 miles (1,600 kilometres), until grease is seen to be coming from each end plate seal. There is one nipple only beneath the centre of the fork bridge.

9 Rear suspension units - removal and refitting

1   Only a limited amount of dismantling can be undertaken because the damper unit is an integral part of each unit and is sealed. If the unit leaks oil or if the damping action is lost, the unit must be replaced as a whole after removing the compression spring and outer shroud.
2   If the units receive attention, place the machine on the centre stand and remove both nuts and bolts so that both units can be removed from the frame. It is possible to remove the suspension units without dismantling the rear mudguard.
3   The spring and outer shield can be removed by clamping the lower end of the unit in a vice and depressing the outer shield so that the semi-circular spring retainers in the top of the shield can be displaced. The outer shield and spring can now be lifted off. The suspension units should be set to their lightest load to make the task easier.
4   When replacing ensure that the rubber bonded bushes in each eye are a tight fit and in good condition. The bushes can be easily renewed by driving out the old one and pressing in the new one, using a smear of soapy water to assist assembly.
5   Squeaking of a suspension unit is normally due to the spring rubbing on the bottom shield. To eliminate the squeaking smear some high melting point grease on the inside of the shield. The plunger rod should not be lubricated.
6   Reassembly is the reversal of dismantling but ensure that the cam is in the light load position before compressing the spring.
7   Variations can occur in dampers and a careful examination should be made of the overall length between the mounting eyes. One damper can be weaker than the other due to weakening of one spring, in which case replace both springs.

10 Rear suspension units - adjusting the loading

1   The rear suspension is controlled by a Girling combined coil spring and hydraulic damper units. The damping mechanism is sealed but the static loading of the spring is adjustable.

2   To adjust first remove the 'C' spanner from the tool kit. Identify the three position castellated cam rings covered by shrouds below the chromium plated springs.
3   Place the machine on the stand to remove as much load as possible and use the 'C' spanner to turn the cams. It is essential that both units are identical in adjustment.
4   Go to the rear of the machine and compare the lengths of the adjustment areas. To increase the static loading turn the castellated cam ring clockwise.

11 Centre stand and spring - overhaul and replacement

1   The centre stand has a three piece spindle. The spindle consists of two half sections threaded at one end, both of which mate with a central spacing section. Each spindle half section has flats machined next to the linking threads. Note also the flatted section on the central spacer.
2   The removal procedure is as follows: First take out the cotter pins in the ends of the half spindles and remove the small springs and washers.
3   Using the flats for spanner leverage, unscrew the two half spindles from the central spacer holding each of the lock nuts on either side of the central spacer in turn with a second spanner. Earlier models did not have these lock nuts fitted. Remove the stand by extracting the half spindles.
4   Clean the stand thoroughly and examine for cracks, damage or rust. A coat of matt black paint will add to its appearance. Pay careful attention to the spring and ensure it is effective in operation.
5   Pay particular attention to the frame attachment lugs for the stand and check for cracks or damage. If the lugs fail then the stand could fall also.
6   The replacement of the stand is the reversal of the removal. Use new cotter pins. A bar or an old screwdriver is needed to lever the spring back into position over its retaining hook. Don't forget to grease the two half spindles, prior to their insertion.

12 Prop stand - examination

1   The prop stand is secured to the frame lug with a single bolt. To remove unscrew the bolt. The return spring will be released as the stand is withdrawn from its lug.
2   Clean the stand thoroughly of grease and road dirt and ex-

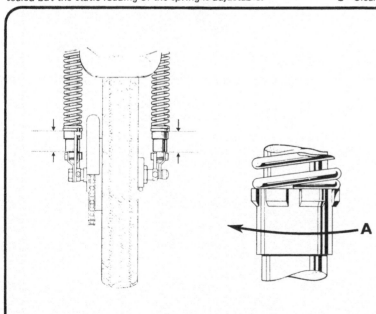

Fig. 6.6. Rear suspension adjustment

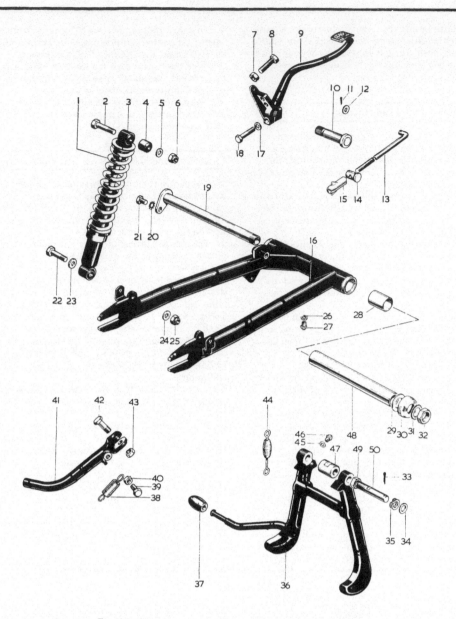

Fig. 6.7. Swinging arm and suspension units (Rocket 3)

1	Rear suspension unit spring - 2 off	26	Washer
2	Bolt - 3 off	27	Grease nipple
3	Rear suspension unit - 2 off	28	Swinging arm bush - 2 off
4	Suspension unit bush	29	Swinging arm thrust bearing - 2 off
5	Washer - 2 off	30	Dust cap - 2 off
6	Nut - 2 off	31	Shakeproof washer
7	Nut - 2 off	32	Nut
8	Bolt - 2 off	33	Split pin - 2 off
9	Rear brake pedal	34	Plain washer - 2 off
10	Brake pedal shaft	35	Spring washer - 2 off
11	Split pin	36	Centre stand
12	Plain washer	37	Rubber
13	Brake rod	38	Spring
14	Pivot pin	39	Prop stand
15	Adjuster nut	40	Bolt
16	Swinging arm with bushes	41	Locknut
17	Washer	42	Bolt
18	Bolt	43	Locknut
19	Swinging arm spindle	44	Spring
20	Washer	45	Washer - 2 off
21	Bolt	46	Grease nipple - 2 off
22	Bolt	47	Coupling
23	Washer - 2 off	48	Swinging arm distance tube
24	Washer - 2 off	49	Nut - 2 off
25	Nut - 2 off	50	Pivot bar - 2 off

amine for cracks, damage, deterioration or rust. A coat of matt black paint is beneficial.

3   Examine the spring for damage or weakness after cleaning thoroughly and replace if at all dissatisfied.

4   Replacement is the reversal of removal. Don't forget to grease the pivot bolt.

13 Removing and refitting the rear mudguard

Triumph

1   Remove the rear wheel as per Chapter 7, Section 9 of this manual.

2   The mudguard blade is secured to the pivot lug bracket by the bottom front bolt nut and washer. Remove these.

3   Release the silencer stay bolt and the bolts securing the hand rail to the rear mudguard. Remove the bolts securing the rear frame clip, support strip and wiring protector. It is necessary to remove the tail clamp housing top nut and plain washer to release the lower end of the wiring protector. Mark all the connections you break for easy refitting.

4   Disconnect and pull the tail lamp wire connectors from beneath the twin seat and through the grommet in the mudguard.

5   To withdraw the mudguard from the frame remove two nuts, bolts and washers securing it to the bridge.

BSA

6   Remove the dual seat and disconnect the rear light cables and take out the rear wheel (see paragraph 1 above).

7   Remove the two frame cross member nuts and bolts at the front of the mudguard. Remove the nuts and bolts securing the mudguard to the bridge piece and the two nuts and bolts through the seat rail bracket.

8   Carefully examine the mudguard for damage or deterioration. Don't be mislaid by the Triumph stores term 'fender', it means the same. Examine the securing points for corrosion, which could mean enlargement of the holes and subsequent sloppiness in attachment. As mudguards are expensive to replace, consult your dealer about a repair if the damage is minor.

9   Reassembly is the reversal of the strip-out but do not overlook any of the electrical connections that must be remade.

14 Oil tank - removal and replacement

BSA

1   Remove the oil tank filter plug and drain the oil. There is a fibre washer to carefully stow when removing the filter plug. Remember to replace the fibre washer when refitting the tank.

2   The seat should be removed as in Section 17 of this Chapter.

3   Detach the oil return pipe from its connection on top of the tank after removing its small fixing clip. Loosen its clip and disconnect the feed pipe underneath the tank.

4   The tank is secured to the dual seat support rail by one bolt and one stud. The bolt attaches to two clips on the top of the tank. These clips are fitted with rubber spacers. The tank is secured at the base by a bolt that passes through a bracket on the frame and into a captive nut in a lug at the base of the tank. This bolt has a rubber sleeve between the tank and the bracket.

5   Disconnect the chain oiler pipe and tank breather and withdraw the tank. The oil tank is rubber mounted to isolate the tank from high frequency vibration which could cause the welded seams to split.

6   Replacement of the tank is the reverse of the removal, but thorough checking of pipes and connections should be made to ensure that oil is not leaking. Replace any suspect pipes or rubber mounting units. This applies to both BSA and Triumph models.

Triumph

7   Drain the oil into a tray. Disconnect the breather pipe and the pipe leading to the right side of the oil cooler.

8   Disconnect the oil feed union nut at the bottom front of the tank and remove the large hexagon headed nut to gain access to remove the oil tank gauze filter.

9   The tank is secured at the top with clips, rubbers and a peg, which can now be removed. Unfasten the base mounting nut and remove the seat retaining wire.

10   It will be necessary to remove the air cleaner.

11   The oil tank should be eased carefully out of the frame after lifting the tank off the bottom spigot.

12   Replacement is the reversal of removal but re-read paragraph 6 of this Section.

15 Instrument nacelle - removal and replacement

1   The speedometer and tachometer are fitted to the instrument nacelle and can be removed complete with this nacelle. The headlamp is first removed.

2   The nacelle itself is secured with four nuts and plain washers which fit on the securing studs but the electrical disconnections listed in paragraph 4 must be made. Refer to the wiring diagram at the end of this manual if in doubt about the reconnecting. It would help to mark the connectors when removing as it will save a lot of time when refitting.

3   Disconnect the drive cables from the two instrument heads. Withdraw the 'Lucar' connections from the ignition switch, light switch and Zener diode heat sink. Disconnect the earth tag from the Zener diode heat sink.

4   Withdraw the snap connectors for the instrument lamps, nacelle warning lights. Disconnect the ground wires at their snap connectors. Withdraw the cut-out button and dip switch and horn push from their snap connectors.

5   In later models the speedometer and tachometer are fitted in brackets and can be removed with these brackets. This was the case with the demonstration model used in our workshop. Earlier versions of the Rocket 3 also had their speedometers and tachometers located in separate brackets.

16 Speedometer and tachometer drive cables - examination and renovation

1   It is advisable to detach the speedometer and tachometer drive cables from time to time, in order to check whether they are adequately lubricated and whether the outer covers are compressed or damaged at any point along their run. A jerky or sluggish movement at the instrument head can often be attributed to a cable fault.

2   To grease the cable, uncouple both ends and withdraw the inner cable. After removing the old grease, clean with a petrol soaked rag and examine the cable for broken strands or other damage.

3   Regrease the cable with high melting point grease, taking care not to grease the last six inches closest to the instrument head. If this precaution is not observed, grease will work into the instrument and immobilise the sensitive movement.

4   If the cable breaks, it is usually possible to renew the inner cable alone, provided the outer cable is not damaged or compressed at any point along its run. Before inserting the new inner cable, it should be greased in accordance with the instructions given in the preceding paragraph. Try and avoid tight bends in the run of a cable because this will accelerate wear and make the instrument movement sluggish.

17 Dual seat - removal and replacement

1   The dual (or twin) seat fitting varies between models. On the BSA version the dualseat is secured by two nuts and washers under the seat at each side of the rear mudguard and sits on four flat rectangular rubber pads. Removal necessitates unhooking the seat at the front after removing the nuts and washers. A rubber buffer is fitted on later models.

2 On the Triumph version two hinges are used to enable the seat to be lifted at right angles. Two small circular rubber buffers on one side of the base. The seat is secured by a knob and catch arrangement. The knob must be pulled to free the seat. A check wire is fitted to control the seat when in the vertical position.

3 BSA twin seats are equipped with 'Quiltop' covers which are retained by sprags in the seat pan. When replacing the plastic cover it must first be soaked in hot water to soften it. With excess water wrung out it can be easily stretched over the seat.

18 Side covers - removal and replacement

1 Side panel removal and replacement can very between models.

2 The BSA side panels are made of steel and are held at three points at each side. The left cover is secured by three 'Oddie' studs. A half turn on the studs will release the cover. Do not forget to replace the cover styling strip after examining the side panel. The right-hand side cover is held by three screws into the oil tank pommels. This cover has oil tank cooling vents and two of the screws also secure the sidecover name plate.

3 The Triumph side panels are fitted as follows: The right panel is secured to the oil tank by Posidriv screws. The left side panel is secured by a plastic knob at the top front and two spigotted rubber mounting bushes at the rear.

19 Petrol tank embellishments - removal

1 Removal of the tank embellishments will vary between BSA and Triumph machines.

2 BSA tank badges on either side are held by slot head screws whilst the Triumph screws are cross head.

3 The knee grips on either machine are glued on like the soles of shoes; do not pull them off unless you have the correct adhesive for replacement.

4 The petrol tank styling strip on BSA machines is secured by four cross head screws complete with washers. The centre band on the Triumph has no screws and is secured by a hook at the front and at the rear is held captive under the rear tank mounting. To remove the strip, remove the nut securing the hook and lift away the strip.

20 Battery and tool carriers - removal and replacement

1 Battery carriers on the Triumph models are almost identical. Although the battery sits on a rubber tray it is extremely advisable to examine the carrier occasionally for spilt acid. A washing with caustic soda and water plus a coating of anticorrosive paint after a spill or overflow, will give added protection. When removing don't lose the spigotted rubber bushes and tubes.

2 To remove the carrier for overhaul proceed as follows:

3 Remove the left side panel and then the fuse and disconnect the battery electrical connections after disconnecting the fuseholder. Unbuckle the rubber battery retaining strap and remove first the battery and then the rubber tray. Remove the long bolts and the nuts at top and bottom to remove the carrier from the frame.

4 After cleaning and examination of the components replacement is the reversal of removal.

5 The tool carrier is removed from the battery carrier by removing two cleveloc nuts, washers and bolts from the top front mounting and the countersunk screw and nut from the base of the carrier. This applies to Triumph machines.

6 Later BSA machines carry a tool box secured to the frame by two bolts. The battery carrier is secured to the frame by three bolts with rubber spacers either side of the three countersunk bolt locations in the base of the tray. Replacement of the BSA battery carrier and tool box is simply the reversal of the removal procedure. Do not omit to refit the strap when refitting the battery, fuse and connections.

7 The Triumph tool kit is supplied in a tool roll for fitting into the tool box. Normally there are nine items. These are:
Ring spanner (1)
'C' spanner (1)
Open ended spanners (3)
Plug spanner (1)
Screwdriver - twin blades (1)
Tommy bar (1)
Tappet spanner (1)
The BSA tool kit is almost identical but replaces one of the open ended spanners with an additional large double ended box spanner.

21 Steering head lock

1 The parking lock is fitted to the centre of the fork top yoke. It is necessary to turn the handlebar to full left lock and turn the key clockwise to lock the steering. Never use undue force.

2 If it is necessary to replace a faulty lock or in the case of a lost key look for a blanking plug in the fork top yoke. Remove the head slug which is hammered into the screw hole to locate a grub screw. Use a small screwdriver with a standard blade to remove the grub screw. The lock can now be lifted clear unless the mechanism is jammed in the locked position when more disassembly may be necessary. The main dealer should be able to replace the key if the number has been kept.

3 A light oil lubrication will help to keep the lock serviceable.

22 Fairing attachment lugs

1 It is possible to mount a fairing to the frame after the headlamp has been removed.

2 Take out the two chromed bolts and washers securing the headlamp. Pull the headlamp away and disconnect the wires at the rear of the lamp.

3 Two lugs are fitted to facilitate mounting a fairing at this point.

23 Cleaning the machine - general

1 After removing all surface dirt with a rag or sponge washed frequently in clean water, the application of car polish or wax will give a good finish to the machine. The plated parts should require only a wipe over with a damp rag, followed by polishing with a dry rag. If, however, corrosion has taken place, which may occur when the roads are salted during the winter, a proprietary chrome cleaner can be used.

2 The polished alloy parts will lose their sheen and oxidise slowly if they are not polished regularly. The sparing use of metal polish or a special polish such as Solvol Autosol will restore the original finish with only a few minutes labour.

3 The machine should be wiped over immediately after it has been used in the wet so that it is not garaged under damp conditions which will encourage rusting and corrosion. Make sure to wipe the chain and if necessary re-oil it to prevent water from entering the rollers and causing harshness with an accompanying rapid rate of wear. Remember there is little chance of water entering the control cables if they are lubricated regularly, as recommended in the Routine Maintenance Section.

24 Fault diagnosis: frame and forks

Symptom	Cause	Remedy
Machine is unduly sensitive to road surface irregularities	Front and/or rear suspension units damping ineffective.	Check oil level in forks. Renew suspension units.
Machine rolls at low speeds	Steering head bearings overtight or damaged	Slacken bearing adjustment. If no improvement, dismantle and inspect head races.
Machine tends to wander. Steering imprecise	Worn swinging arm suspension bearings.	Check and if necessary renew bushes.
Fork action stiff	Fork legs twisted in yokes or bent	Slacken off wheel spindle clamps, yoke pinch bolts and fork top nuts. Pump forks several times before retightening from bottom. Straighten or renew bent forks.
Forks judder when front brake is applied	Worn fork bushes Steering head bearings slack	Strip forks and renew bushes. Re-adjust to take up play.
Wheels seem out of alignment	Frames distorted through accident damage	Strip frame and check for alignment very carefully. If wheel track is out, renew or straighten parts involved. Broken tube will be self-evident.

Chapter 7 Wheels, brakes and tyres

Contents

Specifications

Wheels

Rim size and type - front	WM 2-19
Rim size and type - rear	WM 3-19
Spoke sizes	
Front (inner) timing side (10)	8/10 swg x 4 11/16 U.H. 95° head
Front (inner) timing side (10) inches	0.160 x 0.128 x 4.687
Front (inner) timing side (10) mm	118.0625
Front (outer) timing side (10)	8/10 swg x 4 11/16 U.H. 80° head
Front (outer) timing side (10) inches	0.160 x 0.128 x 4.687
Front (outer) timing side (10) mm	118.0625
Front (outer) drive side (20)	8/10 swg x 5 5/8 U.H. straight
Front (outer) drive side (20) inches	0.160 x 0.128 x 5.625
Front (outer) drive side (20) mm	219.075
Rear - timing side (20)	10G x 8 3/8 U.H. 90°
Rear - timing side (20) inches	0.128 x 8.375
Rear - timing side (20) mm	212.725
Rear - drive side (20)	8 x 10 swg butted 8¾ U.H. 90° head
Rear - drive side (20) inches	0.128 x 8.00
Rear - drive side (20) mm	212.725

Wheel bearings

Front (left and right hand) mm	20 x 47 x 14 Ball journal
Rear (left and right hand) mm	20 x 47 x 14 Ball journal
Spindle diameter (front) inches	0.7862 - 07867
Spindle diameter (front) mm	19.9695 - 19.9822
Spindle diameter (rear) inches	0.7867 - 0.7862
Spindle diameter (rear) mm	19.98 - 19.984

Rear wheel sprocket

Number of teeth	53 (52 before 1970)
Chain size inches	5/8 x 3/8 x 107*

*number may vary slightly, according to machine specification

Chain size mm	...		...	...	...	...	...	...	15.875 x 9.525

Brakes (floating shoes)

Diameter (front) inches	...		..	...	...	...	...	...	8 ± 0.002
Diameter (front) mm	...		...	...	...	...	...	...	203.2 ± 0.0508
Diameter (rear) inches	...		...	...	...	...	...	...	7 ± 0.002
Diameter (rear) inches	...		..	...	...	...	...	...	177.8 ± 0.0508
Width (front) mm	...	...	...	...	...	...	...	...	41.275
Width (rear) inches	...	...	...	...	...	...	...	...	1¼
Width (rear) mm	...	...	...	...	...	...	...	...	31.7
Lining thickness (front) inches		..	...	...	...	...	...	...	0.181 - 0.188
Lining thickness (front) mm	...	...	...	...	...	...	...	...	4.7
Lining thickness (rear) inches		...	...	..	...	...	...	...	0.165 - 0.175
Lining area (front) sq. ins.	...	...	...	...	...	...	...	...	23.4
Lining area (front) sq. cms.	...	....	...	...	...	...	...	...	150.967
Lining area (rear) sq. ins.	...	...	...	...	...	...	...	...	14.6
Lining area (rear) sq. cms.	...	...	...	...	...	...	...	...	94.193

Tyres

									Triumph	BSA
Size (front) inches	...	...	...	...	...	...	...	...	4.10 x 19	3.25 x 19 (K70)
Size (front) mm	...	...	...	...	...	...	...	...	101.1 x 48.2	82.55 x 482
Size (rear) inches	....	...	...	...	...	...	...	...	4.10 x 19	
Size (rear) mm	...	...	...	...	...	...	...	...	101.1 x 48.2	
Pressure (front) lbs/sq ins	...	...	...	...	...	...	...	...	26	
Pressure (front) kgm/sq cm	...	...	...	...	...	...	...	...	1.828	
Pressure (rear) lbs/sq ins	...	...	...	...	...	...	...	...	28	
Pressure (rear) kgm/sq cm	...	...	...	...	...	...	...	...	1.97	

Note: Tyre pressures are for a 76 kg (12 stone) rider. If a pillion passenger is carried increase pressure in both tyres by 6 lbs/sq inch. For sustained speeds over 100 mph pressures should be 32 pounds/sq inch (2.25 kg/sq cm) front and 33 pounds/sq inch (2.32 kg/sq cm) rear.

1 General description

1   The Triumph three cylinder models covered by this manual are now fitted with a front and rear wheel size 19 inches (48.3 cms) rim diameter, with 4.10 inches (10.25 cms) section tyres. This does not apply to earlier models. Earlier Triumph models were fitted with the following tyres:

Front	3.50 x 19 inches
Rear	4.10 x 19 inches

Earlier BSA models were fitted with:

Front	3.25 x 19 (K70)	(82.55 x 482.0 mm)
Rear	4.10 x 19 (K81)	(101.1 x 482.0 mm)

When replacing tyres the size and type required is clearly marked on the tyre and do not change to another size and type unless advised by your dealer. It is possible that there is a cheaper tyre available (with stamps) to fit your machine and even second-hand ones can be obtained but stick to the ones recommended by the manufacturer.

2   The front wheel has an 8 inch (20.3 cm) diameter brake and the rear wheel has a 7 inch (17.8 cm) diameter brake.

3   Later Triumph models have a hydraulic disc brake fitted to the front wheel.

The major components of the disc brake are the brake lever, master cylinder, pressure switch, brake line, caliper assembly and the disc. The pulling of the brake lever by the rider causes a piston to exert pressure on the brake fluid. This pressure is transmitted through the brake line to the caliper assembly. The caliper assembly grips the disc attached to the wheel and stops it.

2 Front wheel - examination and removal

1   Place the machine on the centre stand so that the front wheel is raised clear of the ground. Spin the wheel and check for rim alignment. Small irregularities can be corrected by tightening the spokes in the affected area, although a certain amount of skill is necessary if over-correction is to be avoided. Any 'flats' in the wheel rim should be evident at the same time. These are more difficult to remove with any success and in most cases the wheel will have to be rebuilt on a new rim. Apart from the effect on stability, there is greater risk of damage to the tyre bead and walls if the machine is run with a deformed wheel, especially at high speeds.

2   Check for loose or broken spokes. Tapping the spokes is the best guide to the correctness of tension. A loose spoke will produce a quite different note and should be tightened by turning the nipple in an anticlockwise direction. Always check for run-out by spinning the wheel again.

3   If several spokes require retensioning or there is one that is particularly loose, it is advisable to remove the tyre and tube so that the end of each spoke that projects through the nipple after retensioning can be ground off. If this precaution is not taken, the portion of the spoke that projects may chafe the inner tube and cause a puncture.

4   If it is necessary to remove the front wheel the front brake cable must first be disconnected from the lever on the brake anchor plate. Remove the brake cable spring pin and detach the outer cable end from the cable stop.

5   There are four clamp bolts, two on each of the fork legs. These should be unscrewed, the caps removed and the front wheel withdrawn.

6   The front wheels with disc brakes do not have a front brake cable. It is possible to remove the wheel without removing the disc pads. See Section 4 of this Chapter.

7   Disconnect the front mudguard stays and then four nuts on either fork to release the caps.

3 Front drum brake assembly - examination, renovation and reassembly

1   First to remove the front wheel, disconnect the brake operating cable by removing it from the slotted ferrules at the lower end.

2   Unscrew the eight nuts securing the two spindle caps to the fork legs and slacken the torque stud nuts.

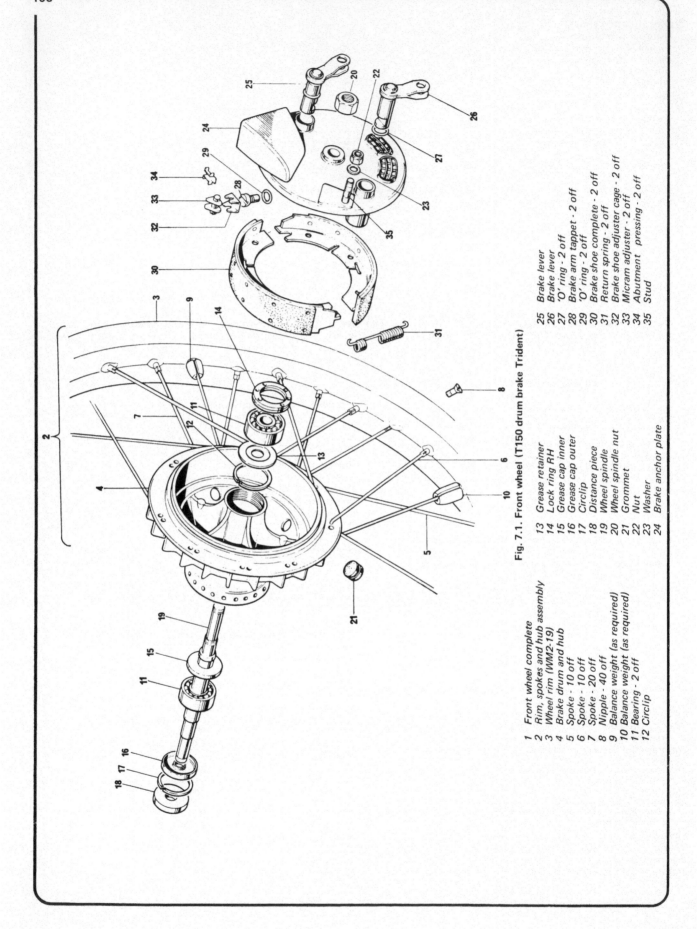

Fig. 7.1. Front wheel (T150 drum brake Trident)

1 Front wheel complete
2 Rim, spokes and hub assembly
3 Wheel rim (WM2-19)
4 Brake drum and hub
5 Spoke - 10 off
6 Spoke - 10 off
7 Spoke - 20 off
8 Nipple - 40 off
9 Balance weight (as required)
10 Balance weight (as required)
11 Bearing - 2 off
12 Circlip

13 Grease retainer
14 Lock ring RH
15 Grease cap inner
16 Grease cap outer
17 Circlip
18 Distance piece
19 Wheel spindle
20 Wheel spindle nut
21 Grommet
22 Nut
23 Washer
24 Brake anchor plate

25 Brake lever
26 Brake lever
27 'O' ring - 2 off
28 Brake arm tappet - 2 off
29 'O' ring - 2 off
30 Brake shoe complete - 2 off
31 Return spring - 2 off
32 Brake shoe adjuster cage - 2 off
33 Micram adjuster - 2 off
34 Abutment pressing - 2 off
35 Stud

3 An anchor plate nut retains the brake plate on the front wheel spindle. When this nut is removed, the brake plate can be drawn away, complete with the brake shoe assembly.

4 Examine the condition of the brake linings. If they are wearing thin or unevenly, the brake shoes should be relined or renewed.

5 To remove the brake shoes from the brake plate, pull them apart whilst lifting them upwards, in the form of a V. When they are clear of the brake plate, the return springs can be removed and the shoes separated. Do not lose the abutment pads fitted to the leading edge of each shoe. (Early BSA/Triumph models).

6 The brake linings are rivetted to the brake shoes and it is easy to remove the old linings by cutting away the soft metal rivets. If the correct Triumph replacements are purchased, the new linings will be supplied ready-drilled with the correct complement of rivets. Keep the lining tight against the shoe throughout the rivetting operation and make sure the rivets are countersunk well below the lining surface. If workshop facilities and experience suggest it would be preferable to obtain replacement shoes, ready lined, costs can be reduced by making use of the Triumph service exchange scheme, available through Triumph agents.

7 Before replacing the brake shoes, check that both brake operating cams are working smoothly and not binding in their pivots. The cams can be removed for cleaning and greasing by unscrewing the nut on each brake operating arm and drawing the arm off, after its position relative to the cam spindle has been marked so that it is replaced in exactly the same position. The spindle and cam can then be pressed out of the housing in the back of the brake plate.

8 Check the inner surface of the brake drum on which the brake shoes bear. The surface should be smooth and free from score marks or indentations, otherwise reduced braking efficiency is inevitable. Remove all traces of brake lining dust and wipe both the brake drum surface and the brake shoes with a clean rag soaked in petrol, to remove any traces of grease. Check that the brake shoes have chamfered ends to prevent pick-up or grab. Check that the brake shoe return springs are in good order and have not weakened.

9 To reassemble the brake shoes on the brake plate, fit the return springs first and force the shoes apart, holding them in a V formation. If they are now located with the operating cams they can usually be snapped into position by pressing downward. Do not use excessive force or the shoes may distort permanently. Make sure the abutment pads are not omitted.

10 A different type of brake unit is fitted to the post 1970 models which have wheels with conical hubs. Although the operating principle is the same, car-type brake shoe expanders are fitted. A micram adjuster is fitted in place of the abutment pads used previously, providing a means of compensating for brake lining wear without having to reduce the angle of the brake operating arms. The brake unit can be dismantled and reassembled by using the procedure already described.

4 Front disc brake assembly - examination, renovation and reassembly

1 The front wheel complete with the brake disc can be withdrawn from the front wheel fork after removing the two clamp bolts retaining each split clamp to the lower fork ends and then withdrawing both clamps. The disc which is secured to the left-hand side of the wheel hub should not require removing unless an examination reveals deep scoring or distortion.

2 The friction pads can be levered out of the brake caliper if the two split pins are first removed. It is unnecessary to remove the wheel to change or check the pads.

3 Renew both pads if there is any doubt about the condition of either one, as they should be renewed in pairs and not singly.

4 Clean out the recesses into which the pads fit and the exposed end of the pistons, using a soft brush. Do not on any account use a solvent cleanser or a wire brush. Finish off by giving the piston faces and the friction pad recesses just a smear of brake fluid.

5 If it is found that the pistons do not move freely or are

seized in position, the caliper is in urgent need of attention and must be removed, drained and overhauled. Seek advice from a professional experienced with motor cycle disc brakes. If a piston is seized, the only satisfactory course of action is renewal of the complete brake caliper unit.

6 To remove the brake caliper unit from the machine, unscrew the union where the hydraulic fluid pipe enters the unit and drain the complete hydraulic system into a clean container. Never reuse brake fluid. The caliper unit can now be detached from its mounting on the lower left-hand fork leg by removing the two retaining bolts.

7 If the caliper unit shows evidence of brake fluid leakage, accompanied by the need to top up the hydraulic fluid reservoir at regular intervals, the piston seals require renewal. This is a simple task which is carried out as follows: Remove the front wheel complete with disc as described in Section 4.1. Lift both friction pads out of position and mark the friction faces so that they are replaced in an identical position. Drain the hydraulic system by placing a clean receptacle below the unit to catch the hydraulic fluid and squeezing the handlebar lever so that both pistons are expelled to release the fluid. Unscrew the caliper end plug which has two peg holes and will require the use of the correct peg spanner tool because it is a tight fit. Remove the two pressure seals from their respective grooves, using a blunt nosed tool to ensure the grooves are not damaged in any way.

8 Wet the new seals with hydraulic fluid and insert the first seal into the innermost bore, making sure it has seated correctly. The diameter of the seal is larger than that of the groove into which it fits, so that a good interference fit is achieved. Furthermore, the sections of the seal and seal groove are different to ensure the sealing edge is proud of the groove. Wet one of the pistons with hydraulic fluid and insert it through the outer cylinder (uncovered by removal of the caliper end plug) so that it passes through into the innermost cylinder bore, closed end protruding approximately 5/16 in. (8 mm) from the mouth of the inner bore.

9 Fit the seal and piston in the outer bore of the caliper unit using an identical procedure. Fit a new 'O' ring seal and replace the end plug, tightening it to a torque setting of 26 lb ft. Replace the friction pads in their original positions after checking that all traces of fluid used to lubricate the various components during assembly have been removed, replace the front wheel and refill the master cylinder reservoir with the correct grade of hydraulic fluid. It will be necessary to bleed the system before the correct brake action is restored by following the procedure described fully in Section 6.

10 Note that all these operations must be carried out under conditions of extreme cleanliness. The brake caliper unit must be cleaned thoroughly before dismantling takes place. If particles of grit or other foreign matter find their way into the hydraulic system there is every chance that they will score the precision made parts and render them inoperative, necessitating expensive replacements.

5 Master cylinder - examination and replacing seals

1 The master cylinder and hydraulic fluid reservoir take the form of a combined unit mounted on the right-hand side of the handlebars, to which the front brake lever is attached. The master cylinder is actuated by the front brake lever and applies hydraulic pressure through the system to operate the front brake when the handlebar lever is manipulated. The master cylinder pressurises the hydraulic fluid in the pipe line which, being incompressible, causes the pistons to move within the brake caliper unit and apply the friction pads to the brake disc. It follows that if the piston seals of the master unit leak, hydraulic fluid will be lost and the braking action rendered much less effective.

2 Before the master unit can be removed and dismantled, the system must be drained. Place a clean container below the brake caliper unit and attach a plastic tube from the bleed screw of the caliper unit to the container. Open the bleed screw one complete turn and drain the system by operating the brake lever until the master cylinder reservoir is empty. Close the bleed screw and remove the pipe.

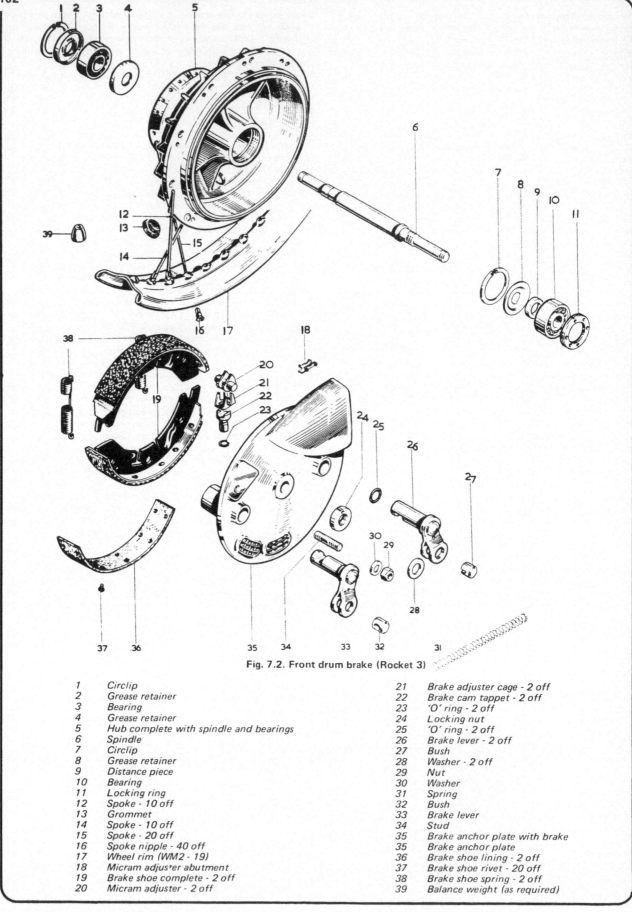

Fig. 7.2. Front drum brake (Rocket 3)

1	Circlip	21	Brake adjuster cage - 2 off
2	Grease retainer	22	Brake cam tappet - 2 off
3	Bearing	23	'O' ring - 2 off
4	Grease retainer	24	Locking nut
5	Hub complete with spindle and bearings	25	'O' ring - 2 off
6	Spindle	26	Brake lever - 2 off
7	Circlip	27	Bush
8	Grease retainer	28	Washer - 2 off
9	Distance piece	29	Nut
10	Bearing	30	Washer
11	Locking ring	31	Spring
12	Spoke - 10 off	32	Bush
13	Grommet	33	Brake lever
14	Spoke - 10 off	34	Stud
15	Spoke - 20 off	35	Brake anchor plate with brake
16	Spoke nipple - 40 off	35	Brake anchor plate
17	Wheel rim (WM2 - 19)	36	Brake shoe lining - 2 off
18	Micram adjuster abutment	37	Brake shoe rivet - 20 off
19	Brake shoe complete - 2 off	38	Brake shoe spring - 2 off
20	Micram adjuster - 2 off	39	Balance weight (as required)

Fig. 7.3. Front brake (disc)

1   Master cylinder
2   Brake lever
3   Piston
4   Spiral pin
5   Barrel and tank assembly
6   Cap
7   Washer
8   Diaphragm
9   Switch housing
10   Trap valve
11   Spring
12   Spring retainer
13   Primary cup
14   Piston washer
15   Piston
16   Secondary cup
17   Circlip
18   Rubber boot
19   Grub screw
20   Pivot screw
21   Nut
22   Screw - 2 off
23   Screw - 2 off
24   Plug
25   Banjo bolt
26   Sealing washer
27   Sealing washer
28   Rubber boot
29   Master cylinder hose
30   Pipe
31   Locknut - 3 off
32   Washer - 3 off
33   Hose
34   Bracket
35   Bolt - 2 off
36   Washer - 2 off
37   Pipe
38   Caliper
39   Brake pad
40   Retaining pin
41   Bleed nipple
42   Piston - 2 off
43   Fluid seal - 2 off
44   Retainer - 2 off
45   Fluid seal
46   Bolt
47   Washer - 2 off
48   Nut - 2 off
49   Caliper cover - 2 off
50   Screw - 2 off
51   Caliper cover decal
52   Screw
Not illus. Bleed nipple cover

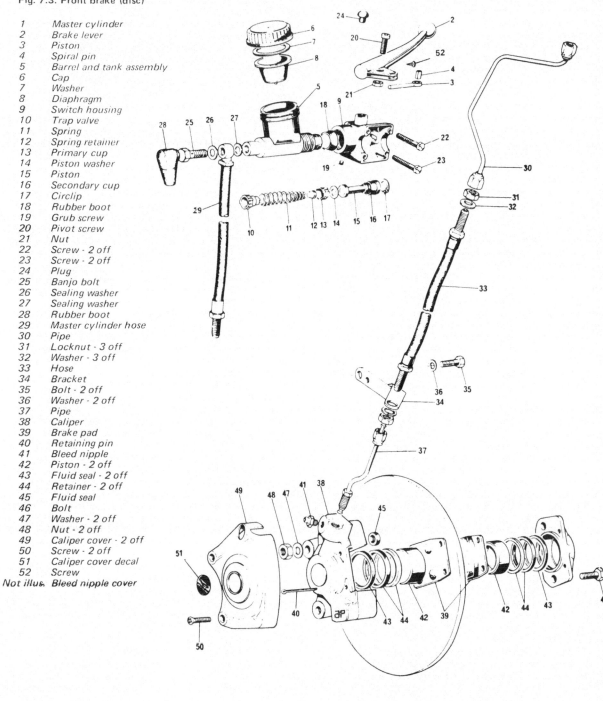

3   To gain access to the master cylinder, disconnect the front brake stop lamp switch by pulling off the spade terminal connections. Lift away the switch cover and detach the hose from the master cylinder by unscrewing the union joint. Remove the four screws securing the unit to the handlebars by means of a split clamp and withdraw the unit complete with integral reservoir.

4   Remove the reservoir cap and bellows seal from the top of the reservoir. Remove the front brake stop lamp switch by unscrewing it from the master cylinder body. Release the handlebar lever by withdrawing the pivot bolt. The rubber boot over the master cylinder piston is retained in position by a special circlip having ten projecting ears. If three or four adjacent ears are lifted progressively, the circlip can gradually be lifted away until it clears the mouth of the bore and is released completely. It will most probably come away with the piston together with the secondary cup.

5   Remove the primary cup washer, cup spreader and bleed valve assembly which will remain within the cylinder bore. They are best displaced by applying gentle air pressure to the hose union bore.

6   Examine the cylinder bore for wear in the form of score marks or surface blemishes. If there is any doubt about the condition of the bore, the master cylinder must be renewed. Check the brake operating lever for pivot bore wear, cracks or fractures, the hose union and switch threads, and the piston for signs of scuffing or wear. Finally, check the brake hose for cuts, cracks or other signs of deterioration.

7   Before replacing the component parts of the master cylinder, wash them all in clean hydraulic fluid and place them in order of assembly on a clean, dust-free surface. Do not wipe them with a fluffy rag; they should be allowed to drain. Particular attention should be given to the replacement primary and secondary cup washers, which must be soaked in hydraulic fluid for at least fifteen minutes to ensure they are supple. Occasional kneading will help in this respect. Clean hands are essential.

8   Commence assembly by placing the unlipped side of the hollow secondary cup over the ground crown of the piston and work it over the crown, then the piston body and shoulder, until it seats in the groove immediately below the piston. Extreme care is needed during the entire operation which must be performed by hand.

9   Fit the boot over the piston, open end toward the secondary cup, and engage the upper end in the piston groove, so that the boot seats squarely.

10 Assemble the trap valve spring over the plastic bobbin; the bobbin must seat squarely in the rubber valve base. Check that the small bleed hole is not obstructed, then insert the plastic spreader in the end of the trap valve spring furthest from the rubber valve. Offer the valve and spring assembly into the master cylinder bore, valve end first, keeping the cylinder bore upright.

11 Insert the primary cup into the bore with the belled-out end innermost. Insert the primary cup washer with the dished portion facing the open end of the bore. Insert the piston, crown end first, into the bore and locate the ten ear circlip over the boot, so that the set of the ears faces away from the cylinder. Apply pressure with a rotary motion and check that the lip of the secondary cup enters the cylinder bore without damaging the lip. Maintain pressure and engage the lower shoulder of the boot with the counter bore of the cylinder that acts as its seating. Work the boot retaining circlip into position, whilst still maintaining pressure on the piston assembly.

12 Still maintaining pressure on the piston assembly, feed the brake lever into position at the fulcrum slot and align the pivot holes so that the pivot bolt can be replaced and locked with the locknut.

6   Front disc brake - bleeding the hydraulic system

1   As mentioned earlier, brake action is impaired or even rendered inoperative if air is introduced into the hydraulic system. This can occur if the seals leak, the reservoir is allowed to run dry or if the system is drained prior to the dismantling of any component part of the system. Even when the system is refilled with hydraulic fluid air pockets will remain and because air will compress, the hydraulic action is lost.

2   Check the fluid content of the reservoir and fill almost to the top. Remember that hydraulic brake fluid is an excellent paint stripper, so beware of spillage, especially near the petrol tank.

3   Place a clean glass jar below the brake caliper unit and attach a clear plastic tube from the caliper bleed screw to the container. Place some clean hydraulic fluid in the container so that the pipe is always immersed below the surface of the fluid.

4   Unscrew the bleed screw one complete turn and pump the handlebar lever slowly. As the fluid is ejected from the bleed screw the level in the reservoir will fall. Take care that the level does not drop too low whilst the operation continues, otherwise air will re-enter the system, necessitating a fresh start.

5   Continue the pumping action with the lever until no further air bubbles emerge from the end of the plastic pipe. Hold the brake lever against the handlebars and tighten the caliper bleed screw. Remove the plastic tube AFTER the bleed screw is closed.

6   Check the brake action for sponginess, which usually denotes there is still air in the system. If the action is spongy, continue the bleeding operation until all traces of air are removed.

7   Bring the reservoir up to the correct level of fluid (within ½ inch of the top of the reservoir) and replace the bellows seal and cap. Check the entire system for leaks. Recheck the brake action.

8   Note that fluid from the container placed below the brake caliper unit whilst the system is bled should not be re-used.

9   When carrying out this exercise, ensure that the machine is firm and level. If the cap is removed when the machine is at an angle then fluid will be lost from the reservoir.

7   Front wheel bearings - removal, examination and replacement

Drum brake models

1   Remove the front wheel from the forks and the brake drum from the wheel.

2   Unscrew the retainer ring from the hub right-hand side using a peg spanner, noting that it has a left-hand thread. The right-hand bearing can be removed by striking the left-hand end of the wheel spindle, thus using its shoulder to drive the bearing out of the hub. Retrieve the grease retainer and, where fitted, the backing ring.

3   The left-hand bearing can be removed using the same method after its circlip has been removed, by inserting the wheel spindle from the right-hand side and driving the bearing out complete with its grease seal and grease retainer. Withdraw the wheel spindle.

4   Wash the bearings in paraffin to remove all old grease and inspect them for wear and damage. Pack the new bearings with grease.

5   Install the grease retainer, followed by the left-hand bearing (use a large socket which bears only on the bearing's outer race, not the balls, to drive it into the hub) and the grease seal. Install the circlip making sure that it engages its groove in the hub. Insert the wheel spindle into the hub right-hand side, shouldered end first, and tap on the end of the spindle so that the grease seal and left-hand bearing are seated against the circlip. Remove the spindle and re-insert it the correct way around.

6   Where fitted, install the backing ring in the hub right-hand side. Install the grease retainer, followed by the bearing, driving it squarely into the hub. Thread the retainer ring securely into place, remembering that it has a left-hand thread. Now tap on the left-hand end of the wheel spindle to bring its shoulder into contact with the inside face of the right-hand bearing.

Disc brake models

7   Remove the front wheel from the forks.

8   Unscrew the wheel spindle fixing nut from the left-hand side. Using a peg spanner unscrew the retainer ring from the hub left-hand side; note that unlike the drum braked models described previously, the retaining ring has a conventional right-hand thread.

9   To remove the left-hand bearing, strike the right-hand end of the wheel spindle until the bearing is freed from the hub. Retrieve the grease retainer. Take care to support the hub whilst the bearing is being driven out to prevent distortion of the brake disc.

10 Before removing the right-hand bearing, remove the circlip from the hub right-hand side. Withdraw the wheel spindle from the hub left-hand side and re-insert it the other way around. Now strike the

1.3 Late models have a hydraulic disc brake

2.6 Front wheel can be removed without disturbing the disc brake assembly

2.7A Disconnect the front mudguard stays

2.7B ... and remove the nuts from the fork leg clamps

4.1 The disc rarely needs to be removed from hub

4.2 Remove split pins to release brake pads

4.3 Remove both pads, never one on its own

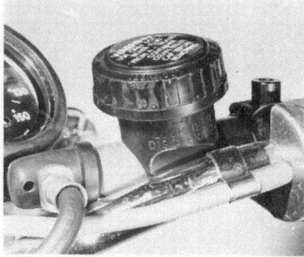
5.1 The combined master cylinder and reservoir assembly

6.9 Never leave cap off reservoir

9.1 Lift chainguard upwards when removing

9.4 Raise rear end of machine to free rear wheel

9.10 Stop lamp switch has spade connectors

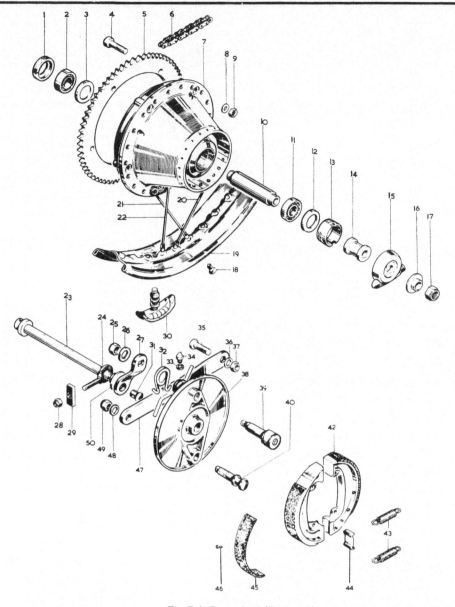

Fig. 7.4. Rear wheel (Rocket 3)

1	Locking ring	26	Washer
2	Bearing	27	Brake lever
3	Support ring	28	Chain adjuster nut - 2 off
4	Bolt - 5 off	29	Chain adjuster end plate
5	Rear wheel sprocket	30	Security bolt - 2 off
6	Rear chain	31	Pivot pin
7	Hub with spindle and bearings	32	Spring
8	Washer - 5 off	33	Washer
9	Nut - 5 off	34	Grease nipple
10	Distance tube	35	Bolt
11	Bearing	36	Washer
12	Grease retainer	37	Nut
13	Speedometer driving ring	38	Brake anchor plate
14	Distance piece	38	Brake anchor plate with brake
15	Speedometer drive gearbox	39	Brake shoe cam
16	Distance piece	40	Brake shoe pivot
17	Nut	41	Leading brake shoe
18	Spoke nipple - 40 off	42	Trailing brake shoe
19	Rim	43	Brake shoe spring - 2 off
20	Spoke - 2o off	44	Brake shoe pivot pad - 2 off
21	Spoke - 10 off	45	Brake lining - 2 off
22	Spoke - 10 off	46	Rivet - 16 off
23	Wheel spindle	47	Torque stay
24	Chain adjuster - 2 off	48	Washer
25	Nut	49	Nut
		50	Distance piece

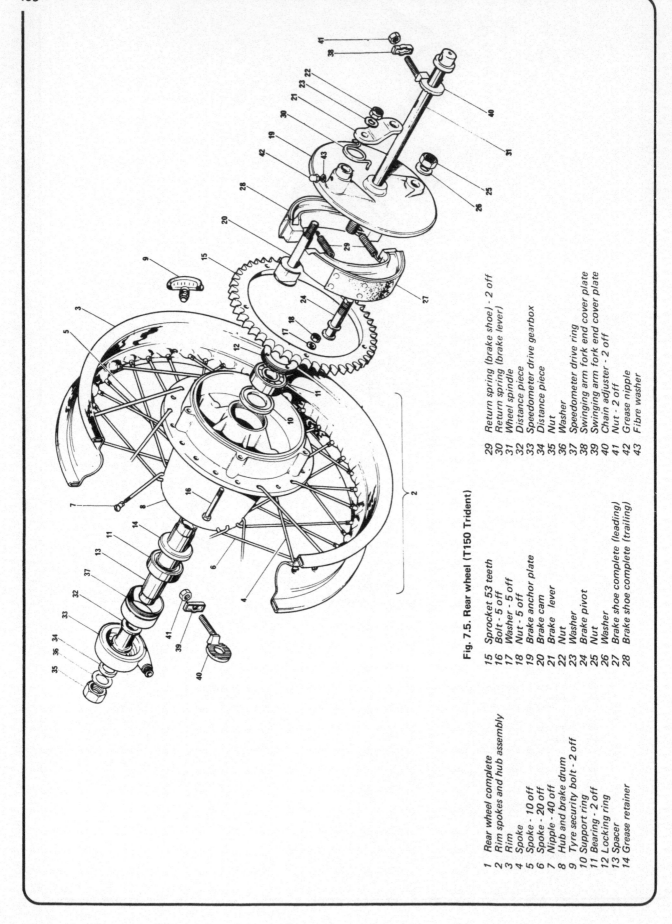

Fig. 7.5. Rear wheel (T150 Trident)

1 Rear wheel complete
2 Rim spokes and hub assembly
3 Rim
4 Spoke
5 Spoke - 10 off
6 Spoke - 20 off
7 Nipple - 40 off
8 Hub and brake drum
9 Tyre security bolt - 2 off
10 Support ring
11 Bearing - 2 off
12 Locking ring
13 Spacer
14 Grease retainer
15 Sprocket 53 teeth
16 Bolt - 5 off
17 Washer - 5 off
18 Nut - 5 off
19 Brake anchor plate
20 Brake cam
21 Brake lever
22 Nut
23 Washer
24 Brake pivot
25 Nut
26 Washer
27 Brake shoe complete (leading)
28 Brake shoe complete (trailing)
29 Return spring (brake shoe) - 2 off
30 Return spring (brake lever)
31 Wheel spindle
32 Distance piece
33 Speedometer drive gearbox
34 Distance piece
35 Nut
36 Washer
37 Speedometer drive ring
38 Swinging arm fork end cover plate
39 Swinging arm fork end cover plate
40 Chain adjuster - 2 off
41 Nut - 2 off
42 Grease nipple
43 Fibre washer

end of the spindle until the bearing is freed from the hub, together with its grease seal and grease retainer. Withdraw the wheel spindle.

11 Wash the bearings in paraffin to remove all old grease and inspect them for wear and damage. Pack the new bearings with grease.

12 Install the grease retainer in the hub right-hand side, followed by the right-hand bearing (use a large socket which bears only on the bearing's outer race, not the balls, to drive it into the hub) and the grease seal. Install the circlip making sure that it engages its groove in the hub. Insert the wheel spindle into the hub left-hand side, shouldered end first, and tap on the end of the spindle so that the grease seal and right-hand bearing are seated against the circlip. Remove the spindle and re-insert it the correct way around.

13 Install the grease retainer in the hub left-hand side, followed by the bearing, driving it squarely into the hub. Thread the retainer ring securely into place. Now tap on the right-hand end of the wheel spindle to bring its shoulder into contact with the inside face of the left-hand bearing. Thread the spindle fixing nut into place and tighten it securely.

8 Front wheel - replacement

1 Replacement of the front wheel is the reverse procedure to the removal.

2 On wheels with drum brakes only fitted, the wheel should be lifted up between the forks and the peg on the right-hand fork leg located in the slot on the brake anchor plate, whilst fitting the spindle ends in the fork bottoms. Pull down on the forks whilst refitting the caps to hold the wheel in position. The spring washers fit under the bolt heads.

3 The right-hand cap bolts are the first to be fully tightened after ascertaining that the brake anchor stop is fully located.

4 Pump the forks up and down to correctly position the left-hand leg and then tighten the left-hand cap bolts. Tighten the bolts to 23 - 25 ft lbs (3.2 - 3.45 kg m).

5 Replace the brake cable and readjust if necessary.

6 Front wheel removal and replacement for wheels fitted with disc brakes is covered in Section 4 of this Chapter. The replacement procedure is the reverse of the removal.

9 Rear wheel - removal and replacement

1 Unscrew the rear brake adjuster and disconnect the rear chain. The bolt at the rear of the chainguard must be slackened so that the chainguard can be swung upwards.

2 Remove the nut securing the rear brake torque stay to the anchor plate. Slacken the left and right wheel spindle securing nuts.

3 Before the rear wheel is removed be sure to disconnect the speedometer cable and disconnect the stop-tail lamp connector.

4 Refitting the rear wheel is the reverse procedure to the removal.

5 Lightly attach the spindle nuts and connect the wheel to the swinging fork.

6 Locate the adjuster caps over the fork ends and lightly tighten the wheel spindle nuts.

7 Fit the chain in position around the rear wheel sprocket and connect up the brake anchor plate torque stay.

8 Before refitting the chain, slacken off both the left and right adjusters.

9 Align the wheels as shown in Section 16 of this Chapter and ensure the spindle nuts and the torque stay securing nuts are correctly tightened.

10 Reconnect the speedometer cable and the stop-tail lamp connection.

10 Rear wheel bearings - removal, examination and replacement

1 The procedure for removing the wheel bearings from the standard rear wheel is identical to that described for the front wheel in Section 7 of this Chapter. In this instance it is necessary to remove the speedometer drive gearbox before access is available

to the right-hand bearing; the gearbox is retained by a locknut and washer. The inner portion of the speedometer drive is threaded into the hub and acts also as the bearing retainer. This too must be removed. Do not lose the outer distance piece between the brake plate and the left-hand bearing.

2 In the case of the quickly detachable wheel, it is necessary to remove the right-hand locknut and detach the speedometer drive gearbox. The inner portion of the speedometer drive must be removed too. This takes the form of a slotted locking ring which threads into the hub and acts as the bearing retainer.

3 If no speedometer drive is fitted, use Triumph tool Z76 to remove the bearing locking ring fitted to the right-hand end of the hub. The ring has a left-hand thread and in the absence of the service tool, can be loosened with a centre punch. The left-hand bearing is preceded by a circlip, which must be removed before the bearing can be displaced. This applies whether or not a speedometer drive gearbox is fitted to the wheel.

4 Because the knock-out wheel spindle has no shoulder, it is necessary to displace the distance piece between the two bearings before the right-hand bearing can be drifted out of position. This is accomplished by inserting a drift from the right-hand side and displacing the distance piece, if necessary by collapsing the inner left-hand grease retainer. The right-hand bearing can then be driven out from the left and vice versa.

5 Follow the procedure given in Section 7, paragraphs 2 and 3, when examining the bearings and replace them by reversing the dismantling procedure. The hub and the bearings must be repacked with new grease prior to assembly.

11 Rear brake - removal and examination

1 Access to the rear brake shoes is obtained as follows:

2 Remove the wheel (see Section 9 of this Chapter).

3 Unscrew the central nut retaining the brake anchor plate.

4 Turn the brake operating lever thus relieving the pressure of the shoes against the drum. This allows the complete brake assembly to be withdrawn from the spindle. Release the lever and remove the return spring. Mark the position of the individual shoes for correct repositioning.

5 Hold the brake plate with one hand and lift up one shoe until it is free. The other shoe can then be removed.

6 Remove the nut and washer securing the brake lever to the cam spindle and remove the lever.

7 Withdraw the cam spindle from the plate.

8 Inspection and overhaul of the brake components is identical to the procedure given in Section 3 of this Chapter for the front drum brake assembly.

9 To reassemble the brake shoes to the anchor plate first place the two brake shoes on the bench in the correct position.

10 Fit the return springs to the retaining hooks with hooked ends uppermost.

11 Take a shoe in each hand at the same time hold the springs in tension.

12 Position the shoes over the cam and fulcrum pin and snap into position. It will be necessary to exert pressure on the outer edges of the shoes.

13 Rotate the brake lever in an anticlockwise direction and engage the return spring.

14 The leading and trailing brake shoes are NOT interchangeable in either the front or rear brake and must be fitted in their correct positions.

12 Rear brake - replacement

1 Reverse the dismantling procedure when replacing the rear brake.

2 Place the reassembled anchor plate over the wheel spindle and secure with the spindle nut.

3 Refitting the rear wheel is covered in Section 9 of this Chapter.

4 Adjustment of the rear brake is covered in Section 18 of this Chapter.

10.5 Pack hub and bearings with grease prior to reassembly

11.3 Rear brake is single leading shoe type

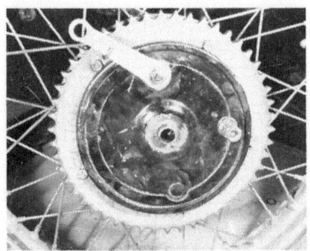

13.1 Inspect rear sprocket for worn teeth

13 Rear wheel sprocket - examination

1   When the rear wheel is being worked on it is advisable to closely inspect the rear sprocket for signs of hooking, erratic wear or broken teeth.
2   If, on checking, any of these defects are evident, the sprocket must be replaced. It is preferable to replace the chain at the same time. Note the rear wheel sprocket is integral with the brake drum.
3   Examine also the gearbox sprocket. A worn rear sprocket is usually an indication that the gearbox sprocket is worn too.

14 Final drive chain - examination and lubrication

1   The rear chain should not be forgotten. All too often it is left until the chain demands attention, by which time it is usually too late.
2   Every 1000 miles, or 500 miles in winter, rotate the wheel to the position where the joining link is positioned on the rear wheel sprocket and slip off the spring link so that the joining link can be pressed out and the chain separated.
3   It is helpful to have an old chain available which can be joined to the existing chain and drawn around the gearbox sprocket until it can be joined with the connecting link. This will free the existing chain for cleaning. Leave the old chain in place and use it to feed the cleaned and greased chain back on when ready.
4   Thoroughly clean the chain with a wire brush and paraffin (Kerosene). Allow it to drain.
5   To check whether the chain needs renewing, lay it lengthwise in a straight line and compress it endwise so that all play is taken up. Anchor one end firmly, then pull endwise in the opposite direction and measure the amount of stretch. If it exceeds ¼ inch per foot, renewal is necessary. Never use an old or worn chain when new sprockets are fitted; it is advisable to renew the chain at the same time so that all new parts run together.
6   Every 2000 miles remove the chain and clean it thoroughly in a bath of paraffin before immersing it in a special chain lubricant such as Linklyfe or Chainguard. These latter types of lubricant are applied in the molten state (the chain is immersed) and therefore achieve much better penetration of the chain links and rollers. Furthermore, the lubricant is less likely to be thrown off when the chain is in motion.
7   When replacing the chain, make sure the spring link is positioned correctly, with the closed end facing the direction of travel. Replacement is made easier if the ends of the chain are pressed into the teeth of the rear wheel sprocket whilst the connecting link is inserted, or a simple 'chain-joiner' is used.

15 Final drive chain - automatic lubrication

1   The rear chain is lubricated automatically by an ingenious metering jet which allows small quantities of oil from an oil tank bleed-off to lubricate the rear chain.
2   If the machine is delivering too much or not enough oil to the chain, a certain amount of trial and error may be necessary before the correct adjustment is achieved.
3   Slacken the locknut located in the neck of the oil tank and adjust the bleed-off as necessary.
4   Turning the adjustment screw clockwise will reduce the flow and vice versa.

16 Adjusting the final drive chain and wheel alignment

1   The rear chain is adjusted by draw bolts fitted to each end of the rear wheel spindle. Correct adjustment is as follows:
2   With the machine on its wheels and the chain at its tightest point the correct adjustment is 0.75 inches (1.8 cm).
3   With the machine on the stand and the chain at its slackest point the correct adjustment is 1.75 inches (4.3 cm).
4   A quick method of checking is with a rider seated on the

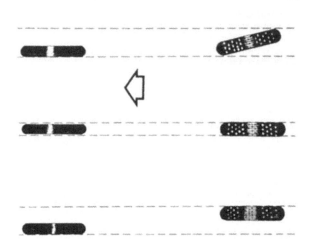

Fig. 7.6. Checking wheel alignment

16.1 The rear chain is adjusted by draw bolts

machine the up and down movement at the tightest point should
be 0.75 inches (1.8 cm).
5   If the adjustment is incorrect, then loosen the wheel spindle
and the nut securing the torque stay to the brake anchor plate and
adjust the draw bolts. Retighten the wheel spindle and recheck
the chain adjustment.
6   If in doubt about the wheel alignment after adjusting the
chain then use a plank or piece of string alongside the rear wheel
and then adjust the draw bolt adjuster so that rear and front
wheels are aligned. If the alignment is not correct the road holding of the machine will be affected and rapid wear on the chain
and rear sprocket will result.
7   When the adjustment is complete check the tightness of the
rear spindle, adjuster draw bolts, brake torque stay nut and then
the adjustment of the brake operating rod.
8   An alternative method of ensuring front and rear wheel alignment is to use two pieces of straight timber about 7 feet long.
9   The machine should be off the stand with the battens placed
alongside the rear wheel on either side and lifted about four
inches from the ground.
10 With both in contact with the rear tyre on both sides the
front wheel should be midway between and parallel to both
battens. This is illustrated in the centre of Fig. 7.6 (checking wheel
alignment). It may be necessary to turn the front wheel a little to
get this alignment.
11 Adjustments can be made to achieve the correct alignment by
turning the rear spindle adjustment nuts. Be sure to maintain the
correct chain adjustment.

18.3 Adjustment is by means of a finger operated nut

17 Front brake - adjustment

Drum brake only
1   The front brake is of the two leading shoe type. The shoe
expansion is equalised by the caliper action of the cam levers. The
control lever at the handlebar combines with the cable adjuster
and these should be adjusted to ensure no slackness in the cable
without applying the brake.
2   If the machine has been much used it may be necessary to re-
position the shoes within the brake drum. Individual adjustment
of the shoes is catered for by a screw at the fulcrum of each shoe.
3   Rotate the front wheel until the adjustment aperture is
opposite the adjuster screw. The adjuster turns with a series of

20.1 Disconnect drive cable by unscrewing gland nut

20.2 Pull the box off the hub. Note spacer within

clicks. Turn it in a clockwise direction to its limit which means that the shoe is fully in contact with the drum. Turn clockwise a couple of clicks until the wheel is free to rotate which means the shoe is just clear of the drum.

4   Turn the wheel through half the circumference and repeat the exercise on the other shoe.

5   Ensure that the wheel revolves freely and readjust the cable length at the handlebar if necessary.

18 Rear brake - adjustment

1   It is possible to adjust the rear brake pedal for position and this should be carried out to the satisfaction of the rider before adjusting the rear brake.

2   From the non-operative position of the brake until the point where the brake bites there should be about ½ inch (1.2 cm) of free movement.

3   Adjustment is by means of a finger operated nut on the rear end of the brake operated rod. To reduce the clearance turn the nut clockwise.

4   Brake adjustment is also necessary when the rear chain alignment has been altered.

5   When the rear brake has been adjusted, check the stop lamp operation and readjust if necessary.

19 Front wheel - balancing

1   It is customary to balance the wheels on high performance machines; an out of balance wheel will be noticed at high speeds by a vibrating wobble transmitted through the handlebars.

2   Set the machine up with the wheels off the ground and spin the front wheel. If it stops in one position each time then the wheel is out of balance. Make sure that the brake is not binding and giving a false reading.

3   Where the wheel stops continuously in one place, the part of the wheel lowest is the heavy part - weights of 1 oz and ½ oz which clamp on the spoke nipple should be positioned opposite the heavy part.

Fig. 7.7a Tyre removal

A   Deflate inner tube and insert lever in close proximity to tyre valve
B   Use two levers to work bead over the edge of rim
C   When first bead is clear of rim, remove tyre as shown

4 Keep adding and removing weights until the wheel can be spun so that it stops in any position. The wheel is now balanced. Make a final check to ensure that the balance weights are secure.

20 Speedometer drive gearbox - general

1 The speedometer drive gearbox is located on the right-hand side of the rear wheel. The drive is by means of a square which accepts the cable end to produce the rotary motion which is required to drive the speedometer head.
2 To take off the drive box take out the rear wheel, unfasten the nut on the outside of the gearbox and pull the box off the hub.
3 When reassembling make sure the slots have properly located with the gearbox drive tabs, before tightening up.

21 Tyres - removal and replacement

1 At some time or other the need will arise to remove and replace the tyres, either as the result of a puncture or because a renewal is required to offset wear. To the inexperienced, tyre changing represents a formidable task yet if a few simple rules are observed and the technique learned the whole operation is surprisingly simple.
2 To remove the tyre from either wheel, first detach the wheel from the machine by following the procedure given in this Chapter whether the front or the rear wheel is involved. Deflate the tyre by removing the valve insert and when it is fully deflated, push the bead of the tyre away from the wheel rim on both sides so that the bead enters the centre well of the rim. Remove the locking cap and push the tyre valve into the tyre.

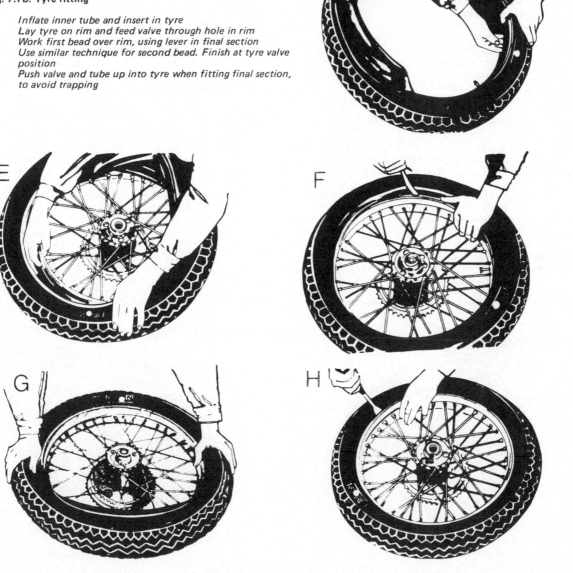

Fig. 7.7b. Tyre fitting

D   Inflate inner tube and insert in tyre
E   Lay tyre on rim and feed valve through hole in rim
F   Work first bead over rim, using lever in final section
G   Use similar technique for second bead. Finish at tyre valve position
H   Push valve and tube up into tyre when fitting final section, to avoid trapping

3   Insert a tyre lever close to the valve and lever the edge of the tyre over the outside of the wheel rim. Very little force should be necessary; if resistance is encountered it is probably due to the fact that the tyre beads have not entered the well of the wheel rim all the way round the tyre.

4   Once the tyre has been edged over the wheel rim, it is easy to work around the wheel rim so that the tyre is completely free on one side. At this stage, the inner tube can be removed.

5   Working from the other side of the wheel, ease the other edge of the tyre over the outside of the wheel rim furthest away. Continue to work around the rim until the tyre is free completely from the rim.

6   If a puncture has necessitated the removal of the tyre, re-inflate the inner tube and immerse it in a bowl of water to trace the source of the leak. Mark its position and deflate the tube. Dry the tube and clean the area around the puncture with a petrol soaked rag. When the surface has dried, apply rubber solution and allow this to dry before removing the backing from a patch and applying the patch to the surface.

7   It is best to use a patch of the self-vulcanising type, which will form a very permanent repair. Note that it may be necessary to remove a protective covering from the top surface of the patch, after it has sealed in position. Inner tubes made from synthetic rubber may require a special type of patch and adhesive if a satisfactory bond is to be achieved.

8   Before replacing the tyre, check the inside to make sure that the agent which caused the puncture is not trapped. Check the outside of the tyre, particularly the tread area, to make sure nothing is trapped that may cause a further puncture.

9   If the inner tube has been patched on a number of past occasions, or if there is a tear or large hole, it is preferable to discard it and fit a new tube. Sudden deflation may cause an accident, particularly if it occurs with the front wheel.

10 To replace the tyre, inflate the inner tube just sufficiently for it to assume a circular shape. Then push it into the tyre so that it is enclosed completely. Lay the tyre on the wheel at an angle and insert the valve through the rim tape and the hole in the wheel rim. Attach the locking cap on the first few threads, sufficient to hold the valve captive in its correct location.

11 Starting at the point furthest from the valve, push the tyre bead over the edge of the wheel rim until it is located in the central well. Continue to work around the tyre in this fashion until the whole of one side of the tyre is on the rim. It may be necessary to use a tyre lever during the final stages.

12 Make sure there is no pull on the tyre valve and again commencing with the area furthest from the valve, ease the other bead of the tyre over the edge of the rim. Finish with the area close to the valve, pushing the valve up into the tyre until the locking cap

touches the rim. This will ensure the inner tube is not trapped when the last section of the bead is edged over the rim with a tyre lever.

13 Check that the inner tube is not trapped at any point. Re-inflate the inner tube and check that the tyre is seating correctly around the wall of the tyre on both sides, which should be equidistant from the wheel rim at all points. If the tyre is unevenly located on the rim, try bouncing the wheel when the tyre is at the recommended pressure. It is probable that one of the beads has not pulled clear of the centre well.

14 Always run the tyres at the recommended pressures and never under or over-inflate. See Specifications for recommended pressures.

15 Tyre replacement is aided by dusting the side walls, particularly in the vicinity of the beads, with a liberal coating of French chalk. Washing up liquid can also be used to good effect, but this has the disadvantage of causing the inner surfaces of the wheel rim to rust.

16 Never replace the inner tube and tyre without the rim tape in position. If this precaution is overlooked there is good chance of the ends of the spoke nipples chafing the inner tube and causing a crop of punctures.

17 Never fit a tyre which has a damaged tread or side walls. Apart from the legal aspects, there is a very great risk of a blow-out, which can have serious consequences on any two wheel vehicle.

18 Tyre valves rarely give trouble but it is always advisable to check whether the valve itself is leaking before removing the tyre. Do not forget to fit the dust cap which forms an effective second seal. This is especially important on a high performance machine, where centrifugal force can cause the valve insert to retract and the tyre to deflate without warning.

22 Security bolt

1   Security bolts are fitted to the rear wheels of high performance machines because the initial take up of drive may cause the tyre to creep around the wheel rim and tear the valve from the inner tube. The security bolt retains the bead of the tyre to the wheel rim and prevents this occurring.

2   There are two security bolts fitted to the rear wheel which are equally spaced either side of the valve in order not to affect the balance of the wheel.

3   Procedures for fitting and removing the tyre remain the same, but in addition the following precautions must be taken.

4   Remove the valve cap and core and then unscrew the security bolts and push them inside the cover.

5   Remove the first bead and then the security bolts from the

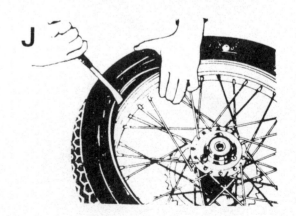

Fig. 7.7c. Tyre security bolts

Fit the security bolt very loosely when one bead of the tyre is fitted

Then fit tyre in normal way. Tighten bolt when tyre is properly seated

rim. Remove the inner tube followed by the second bead and tyre.

6 Replacing the tyre and inner tube goes as follows: Fit the rim tape and then the first bead to the rim without the inner tube. Assemble the bolts on the rim with the nuts on the first few threads.

7 Partly inflate the tube and fit into the tyre.

8 Fit the second bead at the same time keeping the security bolts pressed well into the tyre, taking great care not to trap the inner tube at the edges.

9 Fit the valve stem nut and pump up the tyre. Bounce the wheel in the security bolt area, and tighten the bolt nuts. Do not overtighten otherwise distortion will occur.

23 Fault diagnosis: wheels, brakes and tyres

Symptom	Cause	Remedy
Handlebars oscillate at low speeds	Buckle or flat in wheel rim, most probably front wheel	Check rim alignment by spinning wheel. Correct by retensioning spokes or rebuilding on new rim.
	Tyre not straight on rim	Check tyre alignment.
Machine lacks power and accelerates poorly	Brakes binding	Warm brake drum provides best evidence. Re-adjust brakes.
Brakes grab when applied gently	Ends of brake shoes not chamfered	Chamfer with file.
	Elliptical brake drum	Lightly skim in lathe (specialist attention required).
Front or rear brake feels spongy	Air in hydraulic system (disc brakes only)	Bleed brake.
Brake pull-off sluggish	Brake cam binding in housing	Free and grease.
	Weak brake shoe springs	Renew if springs have not become displaced.
	Sticking pistons in brake caliper (disc brake only)	Overhaul caliper unit.
Harsh transmission	Worn or badly adjusted final drive chain	Adjust or renew as necessary.
	Hooked or badly worn sprockets	Renew as a pair.
	Loose rear sprocket	Check sprocket retaining bolts.

Chapter 8 Electrical system

Contents

Specifications

12 volt Electrical system

Battery	1 Lucas 12v battery PU25A 8 ampere hour
Rectifier	2DS 506 Lucas
Alternator	RM 20 (encapsulated) Lucas
Horns(2)	Windtone P102 R.H. or Clear Hooter* SF 725H -
	Windtone P101 L.H. or Clear Hooter* SF 725L-
	* 1970 models

Bulbs

Headlight	12v 45/40 watt - Pre focus - Lucas 370
Parking light	12v 6 watt M.C.C. - Lucas 989
Stop and tail light	12v 21/6 watt Offset pins - Lucas 380
Speedometer light	12v 2.2 W M.E.S. - Lucas 987
Oil pressure warning light	2w (WL15) - Lucas 281
Hi-beam Warning light	2w (WL15) - Lucas 281
Direction indicator warning light	2w (WL15) - Lucas 281
Flashing indicator lights	12v 21w - Lucas 382
Zener diode type	ZD 715 Lucas
Fuse rating	35A

Battery specific gravity (relative density) U.K. and/or below 90°F (32.2°C)

Filling	1.260
Fully charged	1.280 - 1.300

Tropical climates over 90°F (32.2°C)

Filling...	1.210
Fully charged	1.220 - 1.240

Battery charging rates:

(Maximum permitted electrolyte temperature during charge)

Climates normally below	80°F (27°C) - 100°F (38°C)
Climates normally between...	80° - 100°F (27° -38°C) - 100°F (43°C)
Climates frequently above	100°F (38°C) - 120°F (49°C)

1 General description

1 The electrical current used in these machines is obtained from an alternator driven from the crankshaft and contained in the timing cover. The alternator output is converted into a direct current by a silicon diode bridge rectifier. This current is in turn used to charge a 12 volt 8 ampere hour battery. The charging current is regulated by a Zener diode. The charging rate without the diode at 3000 r.p.m. could go up to 9.5 amperes.

2 This battery operated system supplies the ignition voltage which is controlled by a triple contact breaker directly operated

by the exhaust camshaft. The contact breakers interrupt the supply to three ignition coils. Three condensers, one for each set of contacts are mounted in a waterproof package. A Section in Chapter 5 (Ignition system) will give information on an alternative transistorised ignition set-up.

3   The electrical system contains the conventional lighting facilities, horn etc. Some machines are equipped with an ammeter (Model T150).

4   There is no emergency start facility as even with a discharged battery in circuit there is sufficient voltage to start the machine. There is sufficient initial output from the alternator to commence the ignition cycle.

5   Last but not least, the system is known as 'positive earth' which means that the positive side of the battery goes to earth, which is the metal frame. The negative side of the battery connects to the facilities via a 35A fuse. All earth connections need checking and tightening from time to time.

2   Alternator - testing

1   The alternator driven from the crankshaft supplies an alternating current to the electrical system. In normal operation it should give no trouble. A malfunction in the electrical system shows a need for the services of an electrical specialist, but with the growth of electrical 'knowhow' many owners possess both the instruments needed and the rudimentary knowledge to handle the less complicated faults.

2   If there is a fault indicated by a low battery voltage, check back through the system first ensuring all connections are made and are clean. If there is no output from the rectifier it is probable that the alternator is functioning incorrectly.

3   Output from the three alternator windings can be checked at the disconnect beneath the engine. A suitable lead resistor of about one ohm is needed whilst the voltage is checked by an A.C. voltmeter. Connect the meter (which can be a mini-multimeter able to read A.C. voltages up to 20 volts) with the one ohm lead resistor in parallel with each of the alternator leads in turn and observe the voltmeter readings. Connect one side of the disconnected lead to the one ohm resistor (heavy duty type) and the other side of the resistor to the rectifier earth. The voltages indicated with the engine running at 3000 r.p.m. should be as follows:

5.0 volts A.C.	(green/white and green/black connected)
8.0 volts A.C.	(green/white and green/yellow connected)
10.0 volts A.C.	(green/white, green/yellow and green/black connected)

4   If there is no output at any stage ensure that the chain hasn't damaged the stator leads. Any other complication may mean a new stator or rotor but rarely both.

3   Battery charging procedure and maintenance

1   The battery is located behind the left-hand side panel. With this panel removed, it will be necessary first to undo the strap securing the battery to its carrier. It is then possible to unscrew the terminals to remove the positive and negative connections. Do not lose the vent tube when removing the battery. The battery sits directly into a rubber tray which in turn fits into the battery carrier. The carrier is bolted into the frame at the rear and top. Do not lose the spigotted rubber bushes when removing the carrier.

2   It is possible to monitor the level of the electrolyte through the side of the translucent polythene container. The level line is clearly marked. Never let the acid fall below the level of the top of the plates as sulphation of the plates will occur. Lift the battery out of the carrier so that the coloured filling line can be seen. Add distilled water if the electrolyte level is below the line.

3   The battery top is specially designed to permit venting of gases. With the cover in position the anti-spill filler plugs are sealed in the venting chamber. A vent pipe union on the side of the top is the exit point for unwanted fumes. There are actually two vents in the battery top but one is sealed off. A polythene tube is attached to the vent to lead the fumes, which are corrosive, away from vulnerable sections of the machine.

4   If you renew the battery it is probable you will receive a dry charged one in return. This is a battery that has been initially charged but the acid has been drained for storage and transit. If your dealer has no charging facilities (which is doubtful) proceed as follows:

5   Remove any sealing at the cover and vents and fill with acid of specific gravity, 1.260 in temperate climates, or 1.210 in tropical climates normally above 32.2° centigrade. This acid filling is really the job for a professional as none must be spilt on the person and machine. Plenty of water to dilute the acid is the answer to spillage. If acid splashes into an eye, get the face and eye under a cold water tap immediately. You will probably need first aid treatment. Your dealer will have the apron, goggles and gloves, which are a legal requirement, to do the job more efficiently. After initially filling the battery it should stand for at least an hour as some of the new electrolyte will be absorbed into the plates. After this time you will need to top up to the coloured line with distilled water. Initially, charge the battery for three hours at 1 ampere before fitting to the machine.

6   The average motor cyclist can save money by handling the less complicated side of the machine electrics. Initially however some money must be spent on tools. Fortunately, however, due to energetic competition from abroad the necessary items are still quite reasonable to purchase.

7   To carry out the tasks outlined in paragraph 5 above, you will need:

a) A battery charger. Don't get a large one as you may need to attach it to the machine if charging overnight.

b) An hydrometer. A miniature type is needed for these batteries.

c) A mini-multimeter. This must be able to handle low D.C. and A.C. volts and be used to carry out continuity check on such items as ignition coils.

8   It is essential to guard against a build-up of acid fumes whilst charging the battery as the fumes are highly inflammable. The combination of oxygen and hydrogen can be ignited by a naked flame.

9   If the battery gives all the signs of being fully charged but is unsatisfactory during service the dealer can check its efficiency on a load tester. This is a very robust resistance linked across the terminals and the test can tell an expert immediately if the battery materials are exhausted. Always pay a few extra pence for a battery from an approved battery dealer. A back street dealer may offer to rebuild your battery for you cheaply using the old case etc., but it is false economy.

10 Regularly clean the top of the battery and clean the vent plugs after ensuring the vent holes are clear. Clean off any corrosion from the terminals and lightly smear with mineral jelly (vaseline, not grease).

11 Specific gravity readings vary with temperature and for convenience these are always adjusted to 60° F. The method used is as follows:

12 If the temperature reads above 60° F by 5° add .020 to the hydrometer reading you take to obtain the true specific gravity. Adjust for every 5° or portion of 5° F.

13 If the temperature reads below 60° F by 5° deduct .020 to the hydrometer reading taken to obtain the true specific gravity. Adjust for every 5° or portion of 5° F.

14 The temperature is read from a small thermometer inserted into the electrolyte. It will be necessary to tilt the battery to take the reading.

15 The term 'specific gravity' is new superseded by the term 'relative density' which you may meet in later manuals. It means precisely the same.

16 The expert will fill your battery with dilute sulphuric acid with relative density (specific gravity) of 1.260. Dilute sulphuric acid of S.G. 1.260 is prepared by slowly pouring one part of concentrated sulphuric acid into three parts of distilled water (by

3.1 Remove battery by unbolting battery strap

4.1 Rectifier coils and voltage regulator are housed under the dualseat

5.2 Zener diode is suspended under headlamp. Note finned heat sink

6.1a Turn connector anticlockwise to release

6.1b Bulb can be fitted in one set position only

6.5 Pilot bulb holder is a push fit into reflector shell

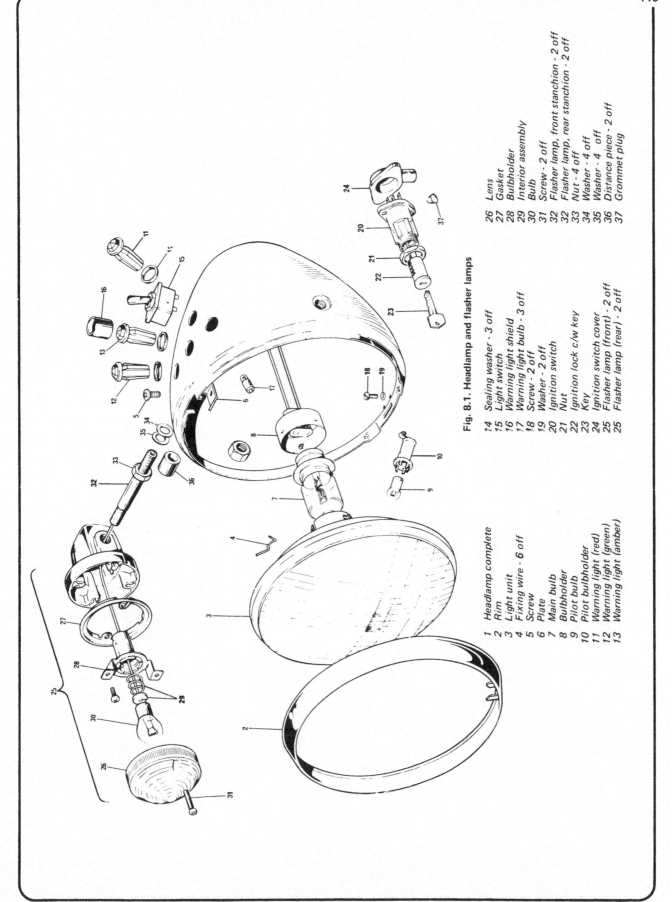

Fig. 8.1. Headlamp and flasher lamps

1   Headlamp complete
2   Rim
3   Light unit
4   Fixing wire - 6 off
5   Screw
6   Plate
7   Main bulb
8   Bulbholder
9   Pilot bulb
10 Pilot bulbholder
11 Warning light (red)
12 Warning light (green)
13 Warning light (amber)

14 Sealing washer - 3 off
15 Light switch
16 Warning light shield
17 Warning light bulb - 3 off
18 Screw - 2 off
19 Washer - 2 off
20 Ignition switch
21 Nut
22 Ignition lock c/w key
23 Key
24 Ignition switch cover
25 Flasher lamp (front) - 2 off
25 Flasher lamp (rear) - 2 off

26 Lens
27 Gasket
28 Bulbholder
29 Interior assembly
30 Bulb
31 Screw - 2 off
32 Flasher lamp, front stanchion - 2 off
32 Flasher lamp, rear stanchion - 2 off
33 Nut - 4 off
34 Washer - 4 off
35 Washer - 4 off
36 Distance piece - 2 off
37 Grommet plug

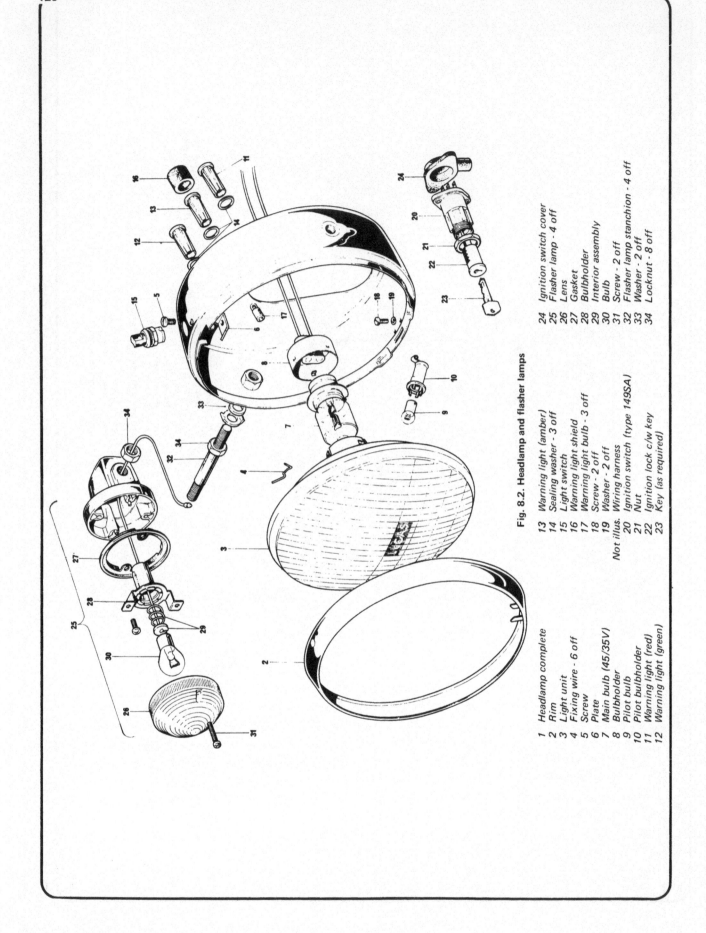

Fig. 8.2. Headlamp and flasher lamps

1 Headlamp complete
2 Rim
3 Light unit
4 Fixing wire - 6 off
5 Screw
6 Plate
7 Main bulb (45/35V)
8 Bulbholder
9 Pilot bulb
10 Pilot bulbholder
11 Warning light (red)
12 Warning light (green)

13 Warning light (amber)
14 Sealing washer - 3 off
15 Light switch
16 Warning light shield
17 Warning light bulb - 3 off
18 Screw - 2 off
19 Washer - 2 off
Not illus. Wiring harness
20 Ignition switch (type 149SA)
21 Nut
22 Ignition lock c/w key
23 Key (as required)

24 Ignition switch cover
25 Flasher lamp - 4 off
26 Lens
27 Gasket
28 Bulbholder
29 Interior assembly
30 Bulb
31 Screw - 2 off
32 Flasher lamp stanchion - 4 off
33 Washer - 2 off
34 Locknut - 8 off

volume). To produce a relative density of 1.260 the mixture is one part of concentrated sulphuric acid into four parts of distilled water. The expert will mix the electrolyte in a glass, earthenware or lead container and allow it to cool. The cells must be filled to the coloured line in one single operation.

4 Silicon diode rectifier - general

1 Charging current for the battery is supplied by the alternator and to convert this alternating current into a suitable direct current for charging use a full wave bridge rectifier is used. The rectifier is made of silicon diode units and is located under the dual seat to the left and rear of the electrical platform.
2 To remove the rectifier remove the nut and then the four electrical connections. The rectifier is of robust construction and is cooled by its own fins. It should give no trouble but if any electrical fault is suspected it must be replaced, not repaired. Check that the connections are clean and that the nut securing the rectifier to the frame is tight. Keep the rectifier clean and clean off any oil that does get onto the exterior immediately.
3 Care should be taken when refitting a metal rectifier as on no account must the plates be twisted out of position. If this occurs, then the rectifying surfaces will be damaged. Hold the nut at the top of the unit firmly with a spanner whilst securing the nut at the base. No excessive force should be necessary.
4 If there is a complicated electrical fault then an auto electrician is needed to check the electrical circuit. Most faults occur due to bad connections, cable damage or fracture. If you have a voltmeter or a multimeter check the battery voltage. With the engine running at about 3,000 r.p.m. the battery voltage should increase due to charging output from the rectifier. Night riding should give an indication of an electrical fault. Switch on the headlamps with the engine stationery. If the lighting is below standard and does not improve with the engine running then there is a fault in the electrical system. This is not necessarily due to the rectifier.
5 To test the rectifier a robust ammeter capable of reading up to 15 amperes D.C. is required. Disconnect the connection at the centre of the rectifier (coloured brown/blue). Make good meter connections between the rectifier centre terminal and the lead you have removed and ensure the ammeter is well insulated from all earthing points and in addition will not slip when the engine is running. At the nominal 3000 r.p.m. which is equivalent to a running speed of 45 m.p.h. then the ammeter should show a charging current of not less than 9.5 amperes. It is advisable to use a centre zero ammeter.

5 Zener diode - general

1 Excessive charge from the alternator is absorbed by a Zener diode which is connected across the battery. It is essential to ensure that the ignition switch is in the 'OFF' position when the machine is not in use as due to the electrical configuration the battery could discharge through the ignition coils. This could cause problems in starting if the ignition coils are over-heating and it would be wise to switch off the ignition and allow the coils to cool.
2 The Zener diode is mounted on a bracket below the headlamp. The bracket is in turn bolted to the middle lug of the fork. The electrical connection to the diode is a brown/white double Lucar connector. Maximum cooling potential is achieved in this location.
3 To replace the diode, remove the electrical connector and then the plastic plug from the top of the heat sink. A nyloc nut secures the diode in position.
4 To clean the heat sink, remove the front bolt from the retaining bracket. It is extremely important to note the connection of the earth wire at this stage. It is coloured double red and is connected at this point. It must NOT be reconnected between the heat sink and diode body.

5 The operation of the Zener diode is as follows: The diode is connected to the centre terminal of the rectifier and is in parallel electrically with the 12 volt battery. If the battery voltage is low then the voltage across the Zener diode will be low and this means the battery will receive all of the rectified alternator output. As the battery is restored to its normal condition the voltage builds up until at approximately 13.5 volts the Zener diode starts to conduct and bleeds off some of the charging current. The excess current fed into the Zener diode causes the diode to get very hot, hence its location at the front of the machine. The heat generated by the excess charge is in turn dissipated by the air stream. From this stage on any increase in battery voltage will increase the Zener diode current. When the battery voltage rises to 15 volts the Zener diode is conducting at a current rate of 5 amperes. If the headlamps are operated this can cause the battery voltage to fall, this in turn reduces the Zener diode current and increases the battery charging current.

The design of the heat sink and its siting at the front of the machine ensures that the diode can if necessary absorb the full output of the alternator.

If it is suspected that the Zener diode is malfunctioning, make a check with the multimeter used in Section 3 of this Chapter. The Zener diode should pass no current at all before 12.75 volts is registered across the battery. If current is passed at that voltage then the Zener diode should be replaced.

6 Headlamp - replacing bulbs and adjusting the beam height

1 To gain access to the bulb and bulb holder it is first necessary to slacken the screws at the top of the headlamp and ease out the rim and light unit. The bulb and its holder can be removed by slight inward pressure coupled with an anticlockwise turn. The new bulb can only be reinstalled in one position to align with the electrical contacts.
2 It is a legal requirement in the United Kingdom that the headlamp be incapable of dazzling any person standing on the same horizontal plane as the machine at a distance greater than twenty five feet from the lamp. The person's eye level is given as not less than three feet six inches above that plane. It is essential therefore that the headlamp be adjusted so that with the rider and pillion passenger seated on the machine, the main beam should be directed dead ahead.
3 To achieve this adjustment choose a flat level area and point the machine towards a wall or garage door. The distance of lamp to wall must be twenty five feet. Chalk a line two feet six inches from the ground on the wall.
4 With rider and pillion passenger seated on the machine the headlamp should be tilted until the beam is focussed at approximately two feet six inches from the base of the wall. The headlamp must be on 'full beam'. The lamp can be tilted by slackening the pivot bolts at either side of the unit. Do not omit to tighten the bolts after the adjustment to a torque of 10 ft lbs (1.4 kg m).
5 The pilot bulb holder is a push fit in the reflector shell. Replacing the pilot lamp or holder is also achieved by lifting out the headlamp rim and light unit as in paragraph 1 of this Section.

7 Tail and stop lamps - replacing bulbs

1 The combined tail and stop lamp unit is fitted with a double filament bulb which has offset pins to prevent incorrect fitting.
2 Access to the bulb is achieved by unscrewing the two slotted screws which secure the lens in position. Do not omit to refit the sealing gasket beneath the lens.
3 Operation of the stop lamp is controlled from the brake pedal and the amount of adjustment is determined by the amount of play at that pedal. Keep the adjuster nuts well greased and periodically clean the operating mechanism. The position of the mechanism can be changed by unloosening the fastening screws and manoeuvring the position of the base to obtain correct brake light operation.

7.1 Remove lens cover for access to tail/stop lamp bulb

10.2 Remove the two screws securing the lens to the lamp body

10.4 The headlamp dip switch forms part of the right-hand handlebar switch assembly

12.1 The direction indicator switch is located on the left handlebar

14.1 The 35 ampere fuse is in the negative battery lead

8 Speedometer and tachometer - replacing bulbs

1   The speedometer and tachometer heads are each fitted with a bulb to illuminate the dial when the headlamp is switched on. The bulb holders are a push fit into the bottom of each instrument case and each carries a 6 watt, 12 volt bulb which has a threaded body.

9 Warning indicators - replacing bulbs and lighting switch

1   With headlamp high beam in use a green warning light illuminates on the binnacle. On some models this is the left of three lamps whilst on others it is the front of two lamps.
2   Illumination of a red lamp indicates lack of oil pressure or that the ignition is on and the engine not functioning. The lamp is operated by the pressure switch screwed into the crankcase below the rear of the oil filter section. The red light operates at pressures below 7 lbs per square inch and should go out with the engine speed exceeding tickover.
3   The warning light bulb holders can be removed only after releasing the headlamp from its brackets. The holders can be pulled down to change the 12 volt, 2 watt bulbs which push fit into the binnacle.
4   The direction indicator warning light is coloured amber and is the centre lamp of the trio. It flashes in conjunction with the main direction indicators. The bulb is replaced in a similar manner to the other warning indicator bulbs.
5   It should be noted that the combination of lamps fitted varies according to the model and year of manufacture.
6   The lighting switch is fitted to the top of the headlamp to the rear of the warning indicators. It is a two position rotary switch. Switch to 'ON' position for headlamp. It is not operative unless the ignition is in position 4.
7   On the T160 model, the indicator lamps form part of a central display that also incorporates the ignition switch. The display is located between the tachometer and speedometer heads.

10 Flashing indicator lamps - location and replacement

1   Flashing indicator lamp units are attached to stanchions fitted on either side of the headlamp body and also at the rear of the machine.
2   To replace the bulbs on either side, remove the two screws securing the lens to the lamp body. The lamp is an bayonet fit into a spring holder.
3   Earlier models of this machine are not equipped with this facility.
4   The direction indicator switch is fitted on the left handlebar. It should be moved down for right indication and up for left indication.

11 Flasher unit - location and replacement

1   The flasher unit is fitted in the electrical component section and to gain access the dual seat must be lifted. The unit is attached to the coil mounting plate.
2   The unit itself is sealed and in case of a fault can only be replaced and not repaired.

12 Headlamp dip switch - location and replacement

1   The headlamp dip switch is located on the left handlebar and it is used to change the headlamp beam between 'main' and 'dipped'.
2   The headlamp flasher switch is also fitted to the left handlebar and is a 'press' switch to flash the headlamp main beam. It is the top switch.
3   The horn push is the lowest switch of the three.
4   In the event of failure of these switches it is necessary to dis-

assemble the complete left-hand handlebar switch assembly.
5   The handlebar passes through the centre of the assembly which is clamped into position with four screws. The switch assembly then divides into two halves, one is the switch half and the other is the lever half (clutch lever section). The switches are not repairable and after ensuring that the electrical connections are not at fault, if a fault persists then the switch half must be replaced.

13 Horn push and horn adjustment

1   A relay is included in the electrical circuit operating the twin windtone horns.
2   The relay is necessary as the operation of the horns requires a large current which if flowing in a long cable from the left handlebar switch could result in a large voltage drop. The handlebar switch therefore operates the relay under the twin seat, the large current passes through the shorter distance of battery - relay - horns.
3   If one or both of the horns fails to operate the following checks and adjustments can be made.
4   A low battery can affect the horns performance, make this the first test.
5   Examine the electrical connections between switch, relay and horns. Clean if necessary.
6   The switch and associated cables can be short circuited by earthing W1 terminal on the relay. If the horns operate and did not before, then the push facility is faulty.
7   The relay can be eliminated by next linking C2 and C1 if this clears the fault then the relay needs replacing.
8   The horns can be adjusted but it is advisable only to operate for short periods due to the large current taken from the battery.
9   Release the plastic protecting cover of each horn by removing the securing screws top and bottom. The adjustment screw is then accessible.
10 Turn the screws a quarter of a turn either way to achieve the best note on each horn. Adjustment is needed only occasionally to compensate for wear of the internal moving parts.
11 Refit the covers and screws after adjustment.

14 Fuse - location and replacement

1   The 35A fuse is in an assembly connected to the negative side of the battery and is in close proximity to that unit.
2   To replace remove any protective covering such as adhesive tape from the fuse holder and holding top and bottom of the fuse holder, press and twist to open the holder. Do not exceed the 35 ampere rating or fit a fuse shorter or wider than the one removed.
3   Always carry at least one spare fuse in the tool kit otherwise your machine can become totally immobilised.
4   In an emergency a piece of silver paper from a cigarette or sweet packet can be lightly wrapped around the fuse and fitted into the fuse holder to provide a link across the fuse holder. This must be removed at the very first opportunity to obtain a new fuse. Make sure the silver paper has a metal content as some paper is 'silvered' with a non-metallic substance that will not conduct.

15 Ignition switch

1   The position of the ignition switch varies with machine model and year. On the demonstration model it is situated on the right side panel. The locks use individual 'Yale' type keys and it is essential to have the key number noted in case of loss of the key and need for a replacement. The key positions are as follows:
2   OFF - Key can be removed. Position vertical.
3   TURN ANTICLOCKWISE - Pilot and tail lights on - key can be removed.

4   TURN CLOCKWISE - Ignition - the engine can be started - the key cannot be removed.

5   NEXT STEP CLOCKWISE - All electrical equipment available for use - the key cannot be removed.

6   In case of a fault on the switch ensure the electrical connectors are correctly located. The switch which on some models is on the headlamp bracket can be replaced by removing the nut securing it in position.

7   Remove the electrical connections before pushing out the switch for replacement.

8   The switch lock has a spring loaded plunger. If it is necessary to replace the switch use a small sharp instrument inserted through a hole in the side of the switch to release the lock to allow removal of lock and switch together.

16 Ignition cut-out ('kill') button

1   The emergency cut-out or 'kill' button is on the right handlebar. It is to stop the engine instantly if the occasion demands. Turn the ignition off in the normal manner after operating this switch.

2   As the switch is rarely used (except when testing) it should require no servicing apart from ensuring the electrical connection is made.

17 Front brake - stop lamp switch

1   In order to comply with traffic requirements in certain overseas countries a stop lamp switch is now incorporated in the front brake cable so that the rear stop lamp is illuminated when the front brake is applied. On models fitted with disc brakes the switch is part of the hydraulic brake reservoir assembly, but can be renewed separately.

2   There is no means of adjustment for the front brake stop lamp switch. If the switch malfunctions the front brake cable must be replaced, or in the case where the switch is incorporated in the master cylinder unit the switch unit must be unscrewed from the master cylinder and renewed.

18 Wiring - layout and examination

1   The cables of the wiring harness are colour coded and will correspond in most models with the wiring diagram in the machine's owners handbook. Typical circuit diagrams are given in this manual.

2   Visual inspection will show whether any breaks or frayed outer coverings are giving rise to short circuits which can cause the main fuse to fail.

3   Snap connectors can cause faults if not correctly and securely located.

4   Short circuits can be traced to chafed wires passing through or close to a frame member. Avoid tight bends to cables especially in the handlebar area. Dim lights can be caused by bad or dirty connections or a partial short circuit on the cable to that light.

19 Fault diagnosis: electrical system

Symptom	Cause	Remedy
Complete electrical failure	Blown fuse	Check wiring and electrical components for short circuit before fitting a new 35 ampere fuse
	Isolated battery	Check battery connections examine for signs of breakage or corrosion
Dim lights and horn inoperative	Battery not charged	Recharge battery with battery charger. Check if generator is giving correct output.
Bulbs constantly blowing	Vibration and/or poor earth connection	Check security of bulb holders - check earth return connections

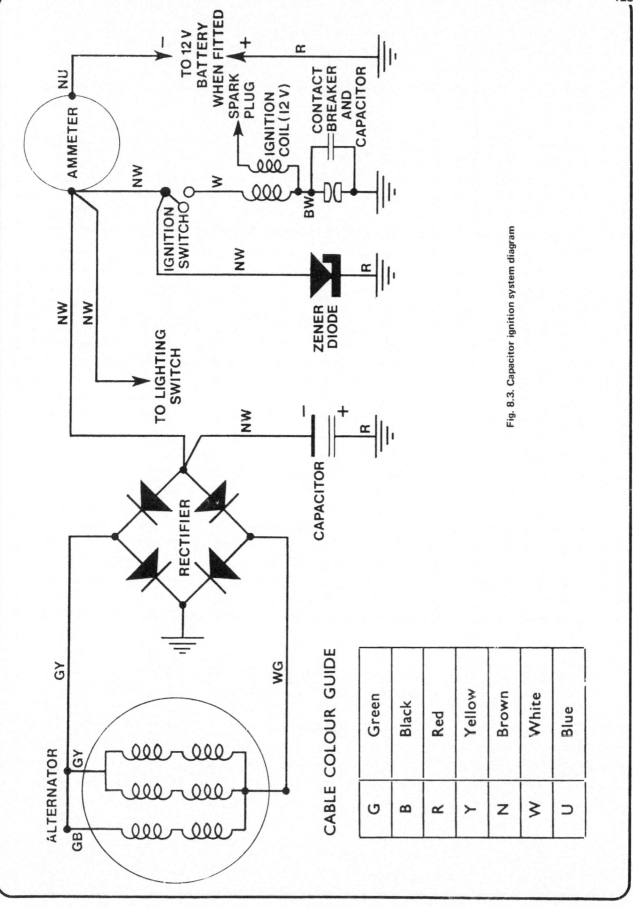

Fig. 8.3. Capacitor ignition system diagram

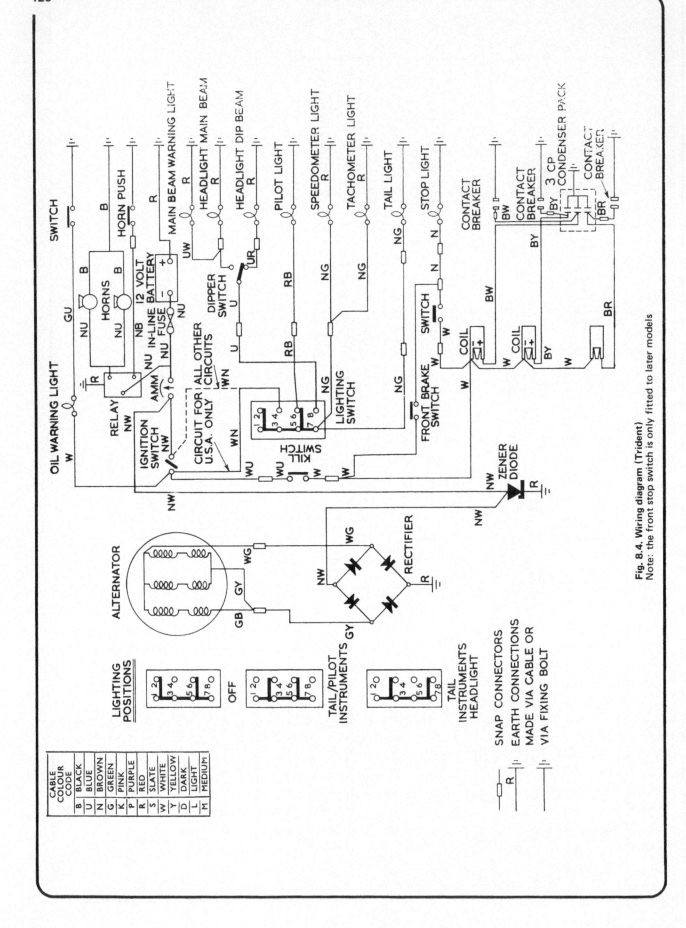

Fig. 8.4. Wiring diagram (Trident)
Note: the front stop switch is only fitted to later models

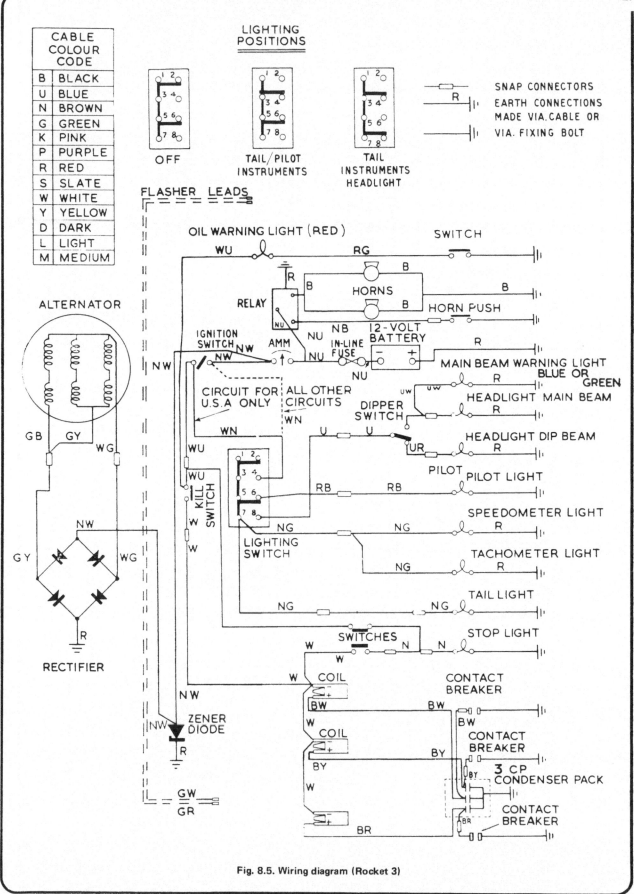

Fig. 8.5. Wiring diagram (Rocket 3)

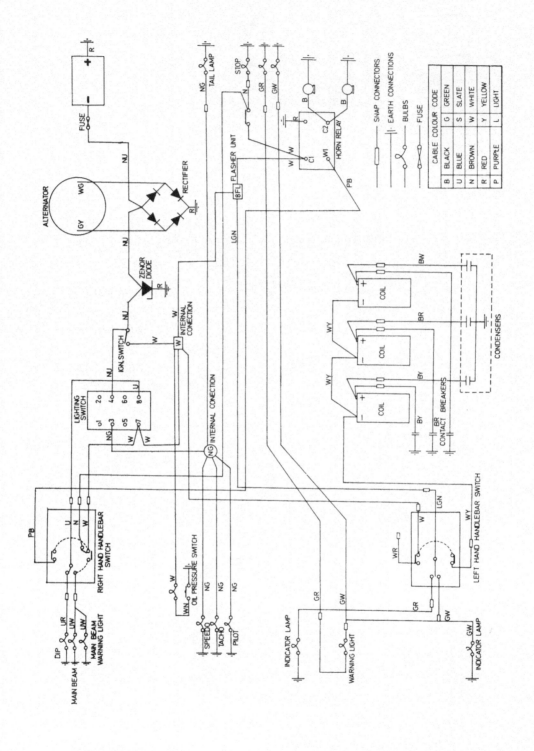

Fig. 8.6. Wiring diagram (T150 model)

Safety first!

Professional motor mechanics are trained in safe working procedures. However enthusiastic you may be about getting on with the job in hand, do take the time to ensure that your safety is not put at risk. A moment's lack of attention can result in an accident, as can failure to observe certain elementary precautions.

There will always be new ways of having accidents, and the following points do not pretend to be a comprehensive list of all dangers; they are intended rather to make you aware of the risks and to encourage a safety-conscious approach to all work you carry out on your vehicle.

Essential DOs and DON'Ts

DON'T start the engine without first ascertaining that the transmission is in neutral.

DON'T suddenly remove the filler cap from a hot cooling system – cover it with a cloth and release the pressure gradually first, or you may get scalded by escaping coolant.

DON'T attempt to drain oil until you are sure it has cooled sufficiently to avoid scalding you.

DON'T grasp any part of the engine, exhaust or silencer without first ascertaining that it is sufficiently cool to avoid burning you.

DON'T allow brake fluid or antifreeze to contact the machine's paintwork or plastic components.

DON'T syphon toxic liquids such as fuel, brake fluid or antifreeze by mouth, or allow them to remain on your skin.

DON'T inhale dust – it may be injurious to health (see *Asbestos* heading).

DON'T allow any spilt oil or grease to remain on the floor – wipe it up straight away, before someone slips on it.

DON'T use ill-fitting spanners or other tools which may slip and cause injury.

DON'T attempt to lift a heavy component which may be beyond your capability – get assistance.

DON'T rush to finish a job, or take unverified short cuts.

DON'T allow children or animals in or around an unattended vehicle.

DON'T inflate a tyre to a pressure above the recommended maximum. Apart from overstressing the carcase and wheel rim, in extreme cases the tyre may blow off forcibly.

DO ensure that the machine is supported securely at all times. This is especially important when the machine is blocked up to aid wheel or fork removal.

DO take care when attempting to slacken a stubborn nut or bolt. It is generally better to pull on a spanner, rather than push, so that if slippage occurs you fall away from the machine rather than on to it.

DO wear eye protection when using power tools such as drill, sander, bench grinder etc.

DO use a barrier cream on your hands prior to undertaking dirty jobs – it will protect your skin from infection as well as making the dirt easier to remove afterwards; but make sure your hands aren't left slippery. Note that long-term contact with used engine oil can be a health hazard.

DO keep loose clothing (cuffs, tie etc) and long hair well out of the way of moving mechanical parts.

DO remove rings, wristwatch etc, before working on the vehicle – especially the electrical system.

DO keep your work area tidy – it is only too easy to fall over articles left lying around.

DO exercise caution when compressing springs for removal or installation. Ensure that the tension is applied and released in a controlled manner, using suitable tools which preclude the possibility of the spring escaping violently.

DO ensure that any lifting tackle used has a safe working load rating adequate for the job.

DO get someone to check periodically that all is well, when working alone on the vehicle.

DO carry out work in a logical sequence and check that everything is correctly assembled and tightened afterwards.

DO remember that your vehicle's safety affects that of yourself and others. If in doubt on any point, get specialist advice.

IF, in spite of following these precautions, you are unfortunate enough to injure yourself, seek medical attention as soon as possible.

Asbestos

Certain friction, insulating, sealing, and other products – such as brake linings, clutch linings, gaskets, etc – contain asbestos. *Extreme care must be taken to avoid inhalation of dust from such products since it is hazardous to health.* If in doubt, assume that they *do* contain asbestos.

Fire

Remember at all times that petrol (gasoline) is highly flammable. Never smoke, or have any kind of naked flame around, when working on the vehicle. But the risk does not end there – a spark caused by an electrical short-circuit, by two metal surfaces contacting each other, by careless use of tools, or even by static electricity built up in your body under certain conditions, can ignite petrol vapour, which in a confined space is highly explosive.

Always disconnect the battery earth (ground) terminal before working on any part of the fuel or electrical system, and never risk spilling fuel on to a hot engine or exhaust.

It is recommended that a fire extinguisher of a type suitable for fuel and electrical fires is kept handy in the garage or workplace at all times. Never try to extinguish a fuel or electrical fire with water.

Note: *Any reference to a 'torch' appearing in this manual should always be taken to mean a hand-held battery-operated electric lamp or flashlight. It does **not** mean a welding/gas torch or blowlamp.*

Fumes

Certain fumes are highly toxic and can quickly cause unconsciousness and even death if inhaled to any extent. Petrol (gasoline) vapour comes into this category, as do the vapours from certain solvents such as trichloroethylene. Any draining or pouring of such volatile fluids should be done in a well ventilated area.

When using cleaning fluids and solvents, read the instructions carefully. Never use materials from unmarked containers – they may give off poisonous vapours.

Never run the engine of a motor vehicle in an enclosed space such as a garage. Exhaust fumes contain carbon monoxide which is extremely poisonous; if you need to run the engine, always do so in the open air or at least have the rear of the vehicle outside the workplace.

The battery

Never cause a spark, or allow a naked light, near the vehicle's battery. It will normally be giving off a certain amount of hydrogen gas, which is highly explosive.

Always disconnect the battery earth (ground) terminal before working on the fuel or electrical systems.

If possible, loosen the filler plugs or cover when charging the battery from an external source. Do not charge at an excessive rate or the battery may burst.

Take care when topping up and when carrying the battery. The acid electrolyte, even when diluted, is very corrosive and should not be allowed to contact the eyes or skin.

If you ever need to prepare electrolyte yourself, always add the acid slowly to the water, and never the other way round. Protect against splashes by wearing rubber gloves and goggles.

Mains electricity

When using an electric power tool, inspection light etc which works from the mains, always ensure that the appliance is correctly connected to its plug and that, where necessary, it is properly earthed (grounded). Do not use such appliances in damp conditions and, again, beware of creating a spark or applying excessive heat in the vicinity of fuel or fuel vapour.

Ignition HT voltage

A severe electric shock can result from touching certain parts of the ignition system, such as the HT leads, when the engine is running or being cranked, particularly if components are damp or the insulation is defective. Where an electronic ignition system is fitted, the HT voltage is much higher and could prove fatal.

Index

Metric conversion tables

Inches	Decimals	Millimetres	Millimetres to Inches		Inches to Millimetres	
			mm	Inches	Inches	mm
1/64	0.015625	0.3969	0.01	0.00039	0.001	0.0254
1/32	0.03125	0.7937	0.02	0.00079	0.002	0.0508
3/64	0.046875	1.1906	0.03	0.00118	0.003	0.0762
1/16	0.0625	1.5875	0.04	0.00157	0.004	0.1016
5/64	0.078125	1.9844	0.05	0.00197	0.005	0.1270
3/32	0.09375	2.3812	0.06	0.00236	0.006	0.1524
7/64	0.109375	2.7781	0.07	0.00276	0.007	0.1778
1/8	0.125	3.1750	0.08	0.00315	0.008	0.2032
9/64	0.140625	3.5719	0.09	0.00354	0.009	0.2286
5/32	0.15625	3.9687	0.1	0.00394	0.01	0.254
11/64	0.171875	4.3656	0.2	0.00787	0.02	0.508
3/16	0.1875	4.7625	0.3	0.01181	0.03	0.762
13/64	0.203125	5.1594	0.4	0.01575	0.04	1.016
7/32	0.21875	5.5562	0.5	0.01969	0.05	1.270
15/64	0.234375	5.9531	0.6	0.02362	0.06	1.524
1/4	0.25	6.3500	0.7	0.02756	0.07	1.778
17/64	0.265625	6.7469	0.8	0.03150	0.08	2.032
9/32	0.28125	7.1437	0.9	0.03543	0.09	2.286
19/64	0.296875	7.5406	1	0.03937	0.1	2.54
5, 16	0.3125	7.9375	2	0.07874	0.2	5.08
21/64	0.328125	8.3344	3	0.11811	0.3	7.62
11/32	0.34375	8.7312	4	0.15748	0.4	10.16
23/64	0.359375	9.1281	5	0.19685	0.5	12.70
3/8	0.375	9.5250	6	0.23622	0.6	15.24
25/64	0.390625	9.9219	7	0.27559	0.7	17.78
13/32	0.40625	10.3187	8	0.31496	0.8	20.32
27/64	0.421875	10.7156	9	0.35433	0.9	22.86
7/16	0.4375	11.1125	10	0.39370	1	25.4
29/64	0.453125	11.5094	11	0.43307	2	50.8
15/32	0.46875	11.9062	12	0.47244	3	76.2
31/64	0.484375	12.3031	13	0.51181	4	101.6
1/2	0.5	12.7000	14	0.55118	5	127.0
33/64	0.515625	13.0969	15	0.59055	6	152.4
17/32	0.53125	13.4937	16	0.62992	7	177.8
35/64	0.546875	13.8906	17	0.66929	8	203.2
9/16	0.5625	14.2875	18	0.70866	9	228.6
37/64	0.578125	14.6844	19	0.74803	10	254.0
19/32	0.59375	15.0812	20	0.78740	11	279.4
39/64	0.609375	15.4781	21	0.82677	12	304.8
5/8	0.625	15.8750	22	0.86614	13	330.2
41/64	0.640625	16.2719	23	0.90551	14	355.6
21/32	0.65625	16.6687	24	0.94488	15	381.0
43/64	0.671875	17.0656	25	0.98425	16	406.4
11/16	0.6875	17.4625	26	1.02362	17	431.8
45/64	0.703125	17.8594	27	1.06299	18	457.2
23/32	0.71875	18.2562	28	1.10236	19	482.6
47/64	0.734375	18.6531	29	1.14173	20	508.0
3/4	0.75	19.0500	30	1.18110	21	533.4
49/64	0.765625	19.4469	31	1.22047	22	558.8
25/32	0.78125	19.8437	32	1.25984	23	584.2
51/64	0.796875	20.2406	33	1.29921	24	609.6
13/16	0.8125	20.6375	34	1.33858	25	635.0
53/64	0.828125	21.0344	35	1.37795	26	660.4
27/32	0.84375	21.4312	36	1.41732	27	685.8
55/64	0.859375	21.8281	37	1.4567	28	711.2
7/8	0.875	22.2250	38	1.4961	29	736.6
57/64	0.890625	22.6219	39	1.5354	30	762.0
29/32	0.90625	23.0187	40	1.5748	31	787.4
59/64	0.921875	23.4156	41	1.6142	32	812.8
15/16	0.9375	23.8125	42	1.6535	33	838.2
61/64	0.953125	24.2094	43	1.6929	34	863.6
31/32	0.96875	24.6062	44	1.7323	35	889.0
63/64	0.984375	25.0031	45	1.7717	36	914.4

1   Imperial gallon = 8 Imp pints = 1.16 US gallons = 277.42 cu in = 4.5459 litres

1   US gallon = 4 US quarts = 0.862 Imp gallon = 231 cu in = 3.785 litres

1   Litre = 0.2199 Imp gallon = 0.2642 US gallon = 61.0253 cu in = 1000 cc

Miles to Kilometres		Kilometres to Miles	
1	1.61	1	0.62
2	3.22	2	1.24
3	4.83	3	1.86
4	6.44	4	2.49
5	8.05	5	3.11
6	9.66	6	3.73
7	11.27	7	4.35
8	12.88	8	4.97
9	14.48	9	5.59
10	16.09	10	6.21
20	32.19	20	12.43
30	48.28	30	18.64
40	64.37	40	24.85
50	80.47	50	31.07
60	96.56	60	37.28
70	112.65	70	43.50
80	128.75	80	49.71
90	144.84	90	55.92
100	160.93	100	62.14

lb f ft to Kg f m		Kg f m to lb f ft		lb f/in^2 : Kg f/cm^2		Kg f/cm^2 : lb f/in^2	
1	0.138	1	7.233	1	0.07	1	14.22
2	0.276	2	14.466	2	0.14	2	28.50
3	0.414	3	21.699	3	0.21	3	42.67
4	0.553	4	28.932	4	0.28	4	56.89
5	0.691	5	36.165	5	0.35	5	71.12
6	0.829	6	43.398	6	0.42	6	85.34
7	0.967	7	50.631	7	0.49	7	99.56
8	1.106	8	57.864	8	0.56	8	113.79
9	1.244	9	65.097	9	0.63	9	128.00
10	1.382	10	72.330	10	0.70	10	142.23
20	2.765	20	144.660	20	1.41	20	284.47
30	4.147	30	216.990	30	2.11	30	426.70

English/American terminology

Because this book has been written in England, British English component names, phrases and spellings have been used throughout. American English usage is quite often different and whereas normally no confusion should occur, a list of equivalent terminology is given below.

English	American	English	American
Air filter	Air cleaner	Number plate	License plate
Alignment (headlamp)	Aim	Output or layshaft	Countershaft
Allen screw/key	Socket screw/wrench	Panniers	Side cases
Anticlockwise	Counterclockwise	Paraffin	Kerosene
Bottom/top gear	Low/high gear	Petrol	Gasoline
Bottom/top yoke	Bottom/top triple clamp	Petrol/fuel tank	Gas tank
Bush	Bushing	Pinking	Pinging
Carburettor	Carburetor	Rear suspension unit	Rear shock absorber
Catch	Latch	Rocker cover	Valve cover
Circlip	Snap ring	Selector	Shifter
Clutch drum	Clutch housing	Self-locking pliers	Vise-grips
Dip switch	Dimmer switch	Side or parking lamp	Parking or auxiliary light
Disulphide	Disulfide	Side or prop stand	Kick stand
Dynamo	DC generator	Silencer	Muffler
Earth	Ground	Spanner	Wrench
End float	End play	Split pin	Cotter pin
Engineer's blue	Machinist's dye	Stanchion	Tube
Exhaust pipe	Header	Sulphuric	Sulfuric
Fault diagnosis	Trouble shooting	Sump	Oil pan
Float chamber	Float bowl	Swinging arm	Swingarm
Footrest	Footpeg	Tab washer	Lock washer
Fuel/petrol tap	Petcock	Top box	Trunk
Gaiter	Boot	Torch	Flashlight
Gearbox	Transmission	Two/four stroke	Two/four cycle
Gearchange	Shift	Tyre	Tire
Gudgeon pin	Wrist/piston pin	Valve collar	Valve retainer
Indicator	Turn signal	Valve collets	Valve cotters
Inlet	Intake	Vice	Vise
Input shaft or mainshaft	Mainshaft	Wheel spindle	Axle
Kickstart	Kickstarter	White spirit	Stoddard solvent
Lower leg	Slider	Windscreen	Windshield
Mudguard	Fender		